Great Careers for People Interested in
Living Things

by
Julie Czerneda

An Imprint of Gale Research Inc.

Copyright © Trifolium Books Inc. and
Weigl Educational Publishers Limited 1993

First published in Canada by Trifolium Books Inc. and
Weigl Educational Publishers Limited

U.S. edition published exclusively by

An imprint of
Gale Research Inc.
835 Penobscot Bldg.
Detroit, MI 48226

Library of Congress Catalog Card Number 93-78080
ISBN 0-8103-9387-5

The activities in this book have been
tested and are safe when carried out
as suggested. The publishers can
accept no responsibility for any
damage caused or sustained by use or
misuse of ideas or materials
mentioned in the activities.

Acknowledgments
The author and the publishers wish to
thank those people whose careers are
featured in this book for allowing us
to interview and photograph them at
work. Their love for their chosen
careers has made our task an easy
one.

Design concept: Julian Cleva
Design and layout: Warren Clark
Editors: Trudy Rising, Jane McNulty
Proofreaders: Diane Klim, Anna Marie Salvia

Printed and bound in Canada
10 9 8 7 6 5 4 3 2 1

This book's text stock contains more
than 50% recycled paper.

Contents

Featured profiles

Careers at a glance

Barbara Mitchell

Equestrian Coach

PERSONAL PROFILE

Career: Equestrian coach. "I'm like a coach in any other professional sport, except I help to train horses as well as people."

Interests: Gardening and downhill skiing.

Latest accomplishment: Taking a horse she trained to the Summer Olympics in Barcelona. "We almost had two riders from our stable at the Olympics. Danny Foster, my partner, was ready to ride, but his horse became injured. Danny's wife Jennifer rode our horse, Zeus, in the equestrian competition."

Why I do what I do: "I love teaching and horses. This job meshes both interests into one."

I am: Outgoing. "I like people a lot."

What I wanted to be when I was in school: "I wasn't sure, but I was interested in science and nature, especially plants."

What an equestrian coach does

"We teach horses to jump fences and to do all the other events in an equestrian competition. We also teach people riding techniques," Barbara explains. Barbara and her partner, Danny Foster, operate a pro-barn. This is a stable where owners keep their horses so that both the rider and the horse learn from an experienced coach like Barbara.

Horse care

"I supervise the care of every horse in the stable," says Barbara. "This includes their exercise schedules, their veterinary care, and their feeding programs. For example, a horse being trained for jumping should have about three times as many oats as a less active horse." Equestrian coaches such as Barbara instruct the grooms employed at a stable. They tell the grooms what each horse needs. Barbara points to a young horse in an outside enclosure, or paddock. "Our horses get turned out every day. Horses are adapted to roaming freely in order to graze. They need time outside to keep physically and mentally fit."

The ride routine

There's a weekly routine at Barbara's pro-barn. "Mondays, we don't ride. But on Tuesdays, we ride the horses on trails through the woods or fields. It's good for them and for us." The rest of the week is spent in the jumping ring. A team of horse and rider may jump every day, or Barbara may prefer that they ease back and work just enough to keep fit.

Four-legged talent

Show-quality horses are rare and valuable. "Showing horses is an international sport — and a business," Barbara notes. Equestrian coaches are often asked to find and buy a horse for a customer. "I act as a talent scout," she says. "I get paid a commission from the purchaser."

Year-round learning

Barbara spends the winter months in the indoor ring. "To begin the day, I have each riding student get a horse ready for a lesson. Then the rider does warm-up work on the lunge rein." The lunge rein allows a person on foot to control a horse, so that it moves in a circle at a certain pace. This helps exercise the horse while teaching it to bend its body as it moves. Barbara coaches riders on how to use hands, legs, and body balance to communicate instructions to the horse. She also works on physical conditioning, helping both human and equine athletes become supple and strong. Part of Barbara's job is to prepare her riders mentally for the challenge of competition.

Great gaits

The walk, trot, and canter are the most common patterns of movement, or gaits, for horses. In a walk, only one foot leaves the ground at a time. This gait is slow, but it helps the horse stay firmly balanced. In a trot, two feet leave the ground at a time. This gait is faster and uses little energy. In a canter or run, all four feet leave the ground for an instant so that the horse can bend its spine for maximum effort.

Mental preparation can mean the difference between winning and losing. "It's vital that the rider knows how to react if the horse does the unexpected. And the rider has to learn how to stay confident, no matter what happens."

When summer arrives, everyone is ready for more advanced work. Barbara asks each rider to go over the jumps. "I sit back, watch the ride, and see exactly how the rider works with the horse," Barbara explains. "Then I know how I can help improve their performance."

"I'm coaching the rider — and the horse. They have to work as a team, both at their best, in order to win," says Barbara.

All in a day's work

Barbara arrives at the barn early in the morning. After helping to feed the horses, she organizes the day to come. "I have very good staff. They know their responsibilities." She makes sure each groom knows about any special care the horses need.

While the horses are turned out into the paddocks, the grooms clean the stalls. "One advantage of being successful at this job," Barbara comments, "is that I only have to muck out stalls once a week!"

The grooms bring the horses back into the barn and give them any treatment or medication they may require. Barbara tells them which horses to get ready for riders. "When the horses are brought back from the lessons, the grooms wash them before putting them into their stalls. This prevents any skin irritation from sweat or dust."

Choosing the tack

Bridles and saddles (called "tack") are carefully fitted to each horse. Horses need different saddles, for example, for show jumping and for dressage. Dressage is an event in which a horse performs specific, often complex maneuvers in response to slight movements of the rider's hands, legs, and weight. It's like a kind of ballet for horses.

During quiet mornings, Barbara sneaks in an hour of bookkeeping or general office work. Once Barbara's students begin to arrive, she's out in the ring. "Regardless of their skill," Barbara notes, "all the people coming here want to improve. They're very dedicated."

Barbara's day fills up quickly. For most of it, she works with her students. But she often spends time with customers as they look over new horses Barbara has bought on their behalf. And if supplies such as a load of wood chips arrive, Barbara takes time out to instruct a crew to distribute it. (The chips are used as bedding in the stalls.)

Barbara selects appropriate tack for each horse.

Phillipa Verry has no arms due to thalidomide poisoning. Nevertheless, Phillipa participates in both dressage and jumping events. She doesn't let her difficulties stop her from doing the things she enjoys.

Horsepower

The term "horsepower" dates from 1783, when steam engines were beginning to replace horses. One horsepower is defined as the amount of work that one typical workhorse can perform in a certain amount of time. The watt (W) has replaced the horsepower as the unit that measures power. One horsepower is equal to 746 W. Think about this: How many "horses" would you need to power an electric handsaw that needs 1560 W?

Winding down

"The grooms are supposed to finish at about 4:00 p.m., but it usually takes everyone a couple more hours to get the horses fed and watered," says Barbara. By 8:00 p.m., Barbara and her husband, John, have found a quiet moment for supper. Maybe.

"I can see the outdoor ring from my kitchen," Barbara confesses. "It's hard for me to stay inside when someone is out riding — especially if it's a beginner who could use some help."

The day's last task is a walk through the barn. "Danny and I water the horses and check that they're comfortable." This is a peaceful time, as the horses are settling and the evening birds begin to sing. Barbara smiles, "I wouldn't miss this moment for anything."

Finally, the lights go off, and everyone rests. "Unless, of course, one of the horses gets sick, or some other problem occurs!" Barbara sighs.

Olympic dreams and heartaches

A horse show is the testing ground for what Barbara does. On show days, Barbara and the horses' grooms travel with the riding students to the competition. Some shows last several days. However, the horses back at the barn are never out of Barbara's mind. "My partner and I try not to be away at the same time."

"Even with a talented rider, a good horse, and an expert coach," Barbara comments, "things can go wrong, and you can lose. Competitive sport can tear your heart out. Sometimes the hardest thing is to hold on to your belief in yourself."

But perseverance and hard work can pay off in the end. "When my partner, Danny, and his horse won the gold medal in show jumping at the Pan-American games in Cuba, it was incredible for all of us. You can't imagine how

Power in the jump

Horses jump well because their muscles and bones are adapted for powerful forward motion. This motion helps wild horses, as well as their relatives like the zebra, to run from danger. The bones of a horse's limbs are very broad, so that a great deal of muscle can be attached to them. However, flexibility is reduced because of this adaptation.

good it felt," Barbara smiles. "Every riding student who comes here dreams of the Olympics."

Activity

A long and rewarding partnership

At what time and place would you have found a living horse that looked like this?

People and horses have been partners for a long time. Make a time chart that shows how people and horses have worked together throughout history. In your chart, include any events that interest you. You may be surprised to learn how important horses were to humans in the past. Your library will have helpful references.

Try to find the following events:
- the first recorded painting of a horse
- the first time horses were used to pull something
- the first time horses were ridden, and by whom
- the last time horses were used in a war in which your country participated
- the last time horses were used for transportation in your town or city

How to become an equestrian coach

Many colleges offer equestrian and stable management programs. This is a fairly recent development. When Barbara started, there weren't many paying jobs involving horses, except at racetracks. "To me, riding and coaching were hobbies." Barbara's interest in horses and nature led her to study science in university.

Barbara also trained as a teacher and taught in high school. "The kids were great, but I couldn't take the four walls," Barbara recalls. "I coached part time as I taught, increasing my knowledge of horses and gaining a reputation for skill and ability. I finally decided to start my own business. More and more people were becoming interested in finding a place where they could entrust their horses to a good coach." The time seemed right to form a partnership with Danny to open a pro-barn. "Our partnership allows each of us to leave the stable to travel to shows and take courses. Having a partner is the way to go," Barbara says.

"Coach certification programs have improved the quality of all equestrian coaching," Barbara explains. These programs are offered by non-profit sports organizations and colleges. "Levels 1 and 2 are similar for every sport. You learn about preparation, sports psychology, and other techniques. At Levels 3 and 4, you specialize in your own sport." Every athlete attending the Olympics must now have a Level 4 coach.

Horse sense

How smart are horses? Most scientists would agree an average horse is not as intelligent as an average dog. Nevertheless, people who work with horses know that they can learn things from watching people. For example, a horse barn never has light switches within reach — most horses quickly learn to turn the lights on and off using their lips. As for how well horses remember things, it's a horse trainer's motto: A horse never forgets a bad experience.

Is this career for you?

Barbara has seen many people come and go within the world of horses and equestrian competition. She knows what sets certain people apart: "Commitment." Horses depend on people for all of their care. Being an equestrian coach is a 365-day-a-year job. "It's something you don't do lightly," Barbara warns. "It's a hard job from the point of view that there's never a time the horses don't need you."

The human side

To be a successful equestrian coach, you have to care about people as well as horses. "Keep in mind that a coach is a teacher," Barbara points out. "You should enjoy your students' achievements as though they were your own." There are rewards, of course. "People who love horses are special to me," Barbara says. "I can go to any country and know I'll be welcome. An interest in horses makes friends in any language." You must be self-confident and willing to accept responsibility, especially if you work for yourself. "Being in charge means there's no one to complain to or to blame," remarks Barbara. "You're it. The responsibility is all yours." Gazing around the barns and the paddocks, she says, "I love being outdoors. This is my idea of an office!"

"Gail is an excellent student," Barbara says proudly. "It's hard to believe she's only been riding here for five months. Her progress comes from both skill and commitment."

Career planning

Attend horse shows or watch them on television. Follow the careers of top horses and riders. Write to any riders who live close to you. "Horse people" will try to help you if they can.

Take a trip to a large stable. Ask people there about their jobs and future plans. If possible, jot notes as you observe employees carrying out their daily tasks.

Making Career Connections

Would you like to know more about horses? If you can't find answers to the questions you have, ask a librarian to help you locate addresses of horse councils and associations. Write to ask specific questions you may have.

Would you like more information on a career as an equestrian coach or on careers that involve work with horses? If so, ask your guidance counselor or a librarian for ideas about organizations to contact for more information.

Getting started

Interested in being an equestrian coach? Here's what you can do now.

1. Ride a lot. Have as much contact with horses, and the people who work with them, as you can. For example, many stables and racetracks invite students to be "hot walkers." A "hot walker" is a person who walks a horse until it is cool, following exercise. Ask for a chance to try this at a stable.

2. Ask at a stable or a pony club if someone wants to share a horse with you. This can be an inexpensive way to have a horse to ride and care for — and the horse benefits from more attention.

3. Keep up your math and science, especially biology and chemistry. You'll need these skills both for business and for horse care.

4. Read as much as you can about horses and how to care for them. Collect magazines and government pamphlets, as well as books.

Related careers

Here are some related careers you may want to check out.

Groom
Usually learns on the job, starting as a hot walker or a stable attendant. Travels to shows or other events with the horses in his or her care.

Large animal veterinarian
After becoming a veterinarian, you can specialize in horse care. Veterinarians may do research on medical problems of horses, or they may give advice on the care of horses.

Stable manager
Responsible for running a stable of horses. This involves hiring grooms, making travel arrangements for the horses, and managing the stable on a day-to-day basis.

Farrier
A specialist in horse-shoeing. Usually self-employed. Must have knowledge of horse anatomy, horse-shoeing techniques, and traditional blacksmithing.

Future watch

"There's a great future in this career," Barbara says firmly. "More people are getting involved with horses all the time. There's a lot of interest in competition and the Olympics." Besides the sport of horse jumping, the recreational enjoyment of horses is increasing. This means that more jobs will be available at every level. "Also, working with horses is a great way for people to be in touch with nature."

Rick Capel

Police Dog Handler

PERSONAL PROFILE

Career: Police officer, Canine Division.

Interests: Being with his family, working out with weights, swimming, and playing squash. "Staying in good shape also helps me do my job."

Latest accomplishment: Completing a university degree in the study of crime.

Why I do what I do: "It's incredibly satisfying and exciting. What I do helps others in a direct way. And I enjoy every minute of working with my dog."

I am: Curious and self-motivated. "I also like making decisions."

What I wanted to be when I was in school: "A police officer. Always."

What a police dog handler does

Rick Capel is a police officer with a difference. He's a dog handler. Rick is the human half of a seven-day-a-week partnership. "My dog is a male German shepherd named Max. He's a veteran of a lot of police work." Rick keeps Max with him all the time he's at work. "Neither one of us can do the job alone." Although they have scheduled working hours, Rick and Max are ready to respond to an emergency at any time.

The dog's job

Police dogs have several roles. For example, they help track suspects and search for drugs or explosives. When officers are faced with a possibly dangerous situation, a police dog can be sent into a building first to find anyone in hiding. Sometimes a police dog is used to catch and hold a suspect who is threatening someone

To keep in top form, the handlers and their dogs get together once a week to practice their skills.

or who is escaping. Max is trained to bite on the arm if ordered. Rick says seriously, "This is a last choice. Fortunately, just the sight and sound of Max are usually enough to convince a suspect to surrender to the police."

The officer's job

As well as enforcing the law, a police dog handler is responsible for the training and care of his or her dog. "My first dog was a 14-month-old German shepherd named Zeus," Rick recalls fondly. "Zeus and I went through training together." Training continues throughout a dog's career. "All the dogs and handlers come in once a week for a training day. If you don't keep up the dogs' skills, they lose them quickly." Because of Rick's experience, he often helps other handlers train their dogs.

Rick looks after Max, by taking him to the veterinarian and by simply playing with him. "One of these days," Rick says, "Max will retire. I'll miss working with him, but I do enjoy looking for new dogs. It's like being a talent scout. Not every dog has the intelligence, health, and temperament for this work. It takes just the right combination. Max here is one of my successes."

Choosing the breed

Different breeds of dogs have different abilities. Springer spaniels, such as the one shown here, like to hunt, and they have a keen sense of smell. So spaniels are used to sniff out explosives or drugs.

Bloodhounds can follow a human scent along the ground, even days later, so they are best for tracking a lost child. German shepherds are a

WANTED
Dogs with the right stuff

Most police dogs are family pets that have grown up to be too difficult for their owners to control properly. Usually when they are under a year old, the dogs are brought to the police. But not all dogs are suited to the Canine Division. "I look for certain things," Rick explains. "First, a strong retrieval drive means the dog will be interested in tracking. I want a dog that's aggressive but not vicious."

Promising dogs go through a two-week trial period with a handler. "At this point some dogs get sent home — usually because we find some quirk in their personalities," Rick says. "I had a dog that was afraid of the dark. It just wouldn't go into a dark room. And another one wouldn't walk on tile floors."

good, all-purpose dog because they are athletic and intelligent. "There are some new kinds of shepherds that are a lot smaller than Max," Rick says. "These compact models certainly have an easier time searching cars!"

All in a day's work

Police officers, including dog handlers, work in eight- to ten-hour shifts: morning, afternoon, and evening. Since most of the "action" happens in the afternoon or evening, these are the most common shifts for Max and Rick.

Working the afternoon shift

Rick and Max arrive at the police station at 4:30 p.m. This gives them half an hour to prepare before beginning their work. "I check the arrest board for arrests made during the past day. Then I look at the wanted posters so I can recognize any suspects," Rick explains. "Officers are also informed of any special concerns, such as outdoor events, for the coming night." Rick grooms Max and then works out in the gym. If necessary, he cleans up remaining paperwork left from the previous day. Then Rick and Max are ready to go.

"Max and I are often the first ones to pursue a fugitive. We've been in a lot of very dangerous situations together and this has forged a special bond between us," says Rick.

"Four on the floor — paws that is!"

"The car I drive is specially designed to carry a dog," Rick explains as he opens the rear door for Max. "The back seat has been taken out and replaced with a fiberglass floor that gives the dog plenty of space." A large bucket of water is built into the floor. "The car is air-conditioned so it won't be too hot for the dog. The windows are tinted. And we only use propane as fuel, so there are no gasoline fumes to bother the dog's sensitive nose."

For the rest of the night, Rick and

A window behind the driver's seat allows Max to leap out of the car if Rick needs him.

Max are on their own. Rick drives around, on the lookout for problems and waiting to respond to calls from other officers. "On a typical night, we get from four to six calls," Rick says. "Of course, it gets busier in the summer."

Anybody home?

One of the most dangerous moments for police officers occurs when they enter a building in search of a suspect who might be armed. Entry becomes much safer when a police dog goes in first. Dogs are trained to search all open rooms. "All the dogs work by scent, but Max is especially good — he even finds people hiding in attics," Rick says. Rick opens any closed doors so Max can search those rooms, too. The instant Max locates someone hiding (or a human scent), he barks as loudly as he can. "Max considers it a game. If he barks, he will get to play with a favorite rubber toy." Rick's voice turns serious. "By going in ahead to locate suspects, Max has saved a lot of lives."

If there are drugs hidden in a car, Max will find them. "He's been trained to do it," Rick explains.

No time to relax

Rick and Max were on the road one evening when a "break-and-enter" call came over the radio. (A break-and-enter occurs when robbers force their way into someone's home or business.) "Max knows something's about to happen when the siren's on and I start picking up the pace. He gets pretty excited."

"When we approach a house on a call like this, there are strict procedures to follow," Rick notes. "We don't approach until backup arrives. Then I go first with Max. This time, Max let me know the instant I opened the door that someone was still inside."

It was the owner, who had been stabbed during the robbery. Other officers attended to him while Rick and Max went into the backyard to try to track the suspect. "Right away Max headed down a ravine behind the house. I could tell he had a good scent by how hard he pulled me after him!" Max quickly led Rick to a stolen VCR, then to a duffel bag, and finally to a jacket — all thrown away by the fugitive. "Max even found the robber's driver's license which had fallen on the ground. I put it into a plastic evidence bag to preserve any fingerprints." The arrest was straightforward after that.

This is the most common work Max does — entering a building first to locate any suspects that might be hiding from police.

Home at last — maybe

At the end of their shift, Rick and Max report back to the station. Rick handles any urgent paperwork, such as information other officers will need immediately. He leaves routine paperwork for the next day. "It's 3:00 a.m. And time to go home. I take Max out for a run and play with him for a while." Max spends his off-duty time in his kennel in Rick's backyard. "Last, but not least, it's time for supper. I feed Max in his pen." Max gets a high-energy diet to keep him fit and alert. "Max knows it's time to rest," Rick says affectionately, "but he's always ready for action." That's just as well, because Rick and Max are often called in to help other officers, even after their shift ends. Rick shrugs. "It doesn't matter what you're doing," he says, "when an unexpected call comes in, you get 'pumped up' right away. That's the kind of work it is."

Activity

The case of the...

This is a report form like the one Rick fills out for every call that he and Max answer.

Part of a dog handler's job is writing a report that tells what happened and how the dog responded during each call. (A call is when an officer is "called" to the scene of a crime.) Imagine you are a police dog handler. You and your dog have been called to help investigate a crime. Write a report, or a short story, that describes what happened.

OAK PARK POLICE DEPARTMENT DOG SEARCH REPORT

Reporting Officer: _____ Dog: _____

Date: _____ Time: _____

Location: _____

Weather Conditions: _____

Suspect

Name: _____ Other Information: _____

Address: _____

Phone #: _____

Suspect

Name: _____ Other Information: _____

Address: _____

Phone #: _____

Details of report _____

If you need ideas, start with one of the calls listed below.
- An impaired (drunk) driver has fled the scene of a car accident.
- A child is missing.
- A bomb threat has been made at an airport.
- Two armed robbers are hidden in a warehouse after injuring a police officer who surprised them.

How to become a police dog handler

Rick knew in high school that he wanted to be a police officer. He went to college and took a two-year law enforcement course. "I worked hard. I knew I'd need to do well to be accepted." After college, Rick applied and was accepted into the police department. "More school!" Rick remembers. "The first step is three weeks of recruit testing, followed by 12 weeks at the police college. Finally, there are six months of riding in a patrol car with an experienced officer, to learn firsthand what happens on the street."

Within the force, officers are promoted as they gain experience and training. "It took me three years to work my way to the top as a uniformed officer." During this time, Rick met a dog handler and decided this was the kind of police work that interested him most. "I'd always liked dogs. The more I found out about working with them, the more I wanted to try it."

Tests and stress

"I had to go through interviews and tests," Rick recalls, "including tests to see how I reacted to sudden life-threatening stress!" Rick was accepted and a dog, Zeus, was assigned to him. Rick and Zeus trained and worked together for four years, earning several awards. "Unfortunately, Zeus developed a condition called hip dysplasia and had to retire." Rick's voice is sad.

Rick stayed with the Canine Division for five years, working with two other dogs, before being transferred to another area when his dogs retired. "Transfers are common within the police department," Rick says. "By working in different departments, you learn a lot more. It makes you a better officer."

Back to the dogs

The Canine Division was Rick's favorite duty. When Max, a dog that Rick had originally trained, came into active duty, Rick was asked to come with him. He didn't hesitate. "This work suits me," Rick says simply. "It's exciting and rewarding. I use all my police training. I help people and I can make a real difference." Rick smiles, stroking Max as he speaks. "And there's the bonus of working with a very special partner."

My kingdom for a horse

Horses as well as dogs have a special place in police work. Trained not to react to sudden noise, police horses are vital in crowd control. This mounted police officer is wearing ceremonial attire — not her on-the-job uniform.

Is this career for you?

If you're thinking about a career as a dog handler, you might be interested in what Rick has to say. "I've seen a lot of people doing this job. The best ones are those who like getting involved. You can't be a person who sits back and lets other people go first."

As he speaks, Rick counts two more characteristics on his fingers. "You need dedication. And you'd better be physically fit!"

Rick also suggests that you assess your own ability to "read" dogs. "Reading" means interpreting the dog's behavior. "For example, Max will dig and bark at a car dashboard if he scents a drug hidden inside. I'm so used to Max that I can tell almost immediately what he's found," Rick chuckles, then becomes more serious. "Reading a dog is a skill some people learn easily and other people never grasp. It really sorts out who will become a handler.

I've learned to have confidence in what Max 'tells' me, especially in life-threatening situations."

Care and trust

You also need something less easy to measure. "You have to be the kind of person who gets emotionally involved," Rick states simply. "You have to be able to care — all the time — about the work you do and about your dog.

"What don't I like about this job? That's a hard question," Rick laughs. "I like everything — absolutely everything — about this type of work." He thinks for a moment. "Well, maybe not *everything*. I really hate it when Max gets sick. You have to understand that I rely on Max completely. He's my bread and butter, but it's more than that. There's a lot of trust and caring between us."

Career planning

Making Career Connections

Check to see if your local police department has a Canine Division. If so, request permission to interview one of the officers involved in the program, conduct the interview, and record your conversation.

Ask an official in your local police department if you can watch some procedures, such as dog training and fingerprinting.

Visit a local dog-training session, and record some of the tips offered in teaching obedience to dogs. Obtain more information about dogs and how to work with them by writing to an organization suggested at the dog-training session.

Write to a local police association or academy for information on a career in police work. Your school guidance counselor or a librarian can help you locate addresses and phone numbers of appropriate organizations to contact.

Getting started

Interested in being a police dog handler? Here's what you can do now.

1. Practice skills that are helpful to police officers, such as writing clear notes, being observant, and taking First Aid courses.

2. Many police forces offer cadet programs for high school students. These can involve a day to several weeks of actual police work experience. Find out if you can participate in one of these programs. Mention your interest in working with dogs.

3. Math and science are essential for criminal investigation work. Plan to take these courses throughout high school. You'll also want to do well in courses that deal with how people behave, such as social studies and family studies.

Related careers

Here are some related careers you may want to check out.

Rescue worker
Trains and uses dogs to find victims of natural disasters, such as earthquakes or avalanches.

Animal trainer
Trains companion animals for the disabled, operates businesses such as obedience training schools for pets, and prepares animal actors for their roles in movies or TV commercials.

Kennel operator/dog breeder
Kennel operators run facilities for keeping dogs overnight. Dog breeders raise dogs for specific purposes, for example, as pets or for shows.

Mounted police officer
Rides a horse on patrol. Responsible for horse care and training. Assists in crowd control and at special events. Uses a horse to work in areas not accessible to cars.

Future watch

The career prospects for dog handlers are excellent. "We're only just beginning to realize how helpful a dog can be," Rick says. For example, dogs have now been trained to signal their handlers when they smell flammable liquids. This allows forensic specialists to determine if a fire might have been started deliberately. Most police forces are planning to add a Canine Division, if they haven't added one already.

Zoo Biologist

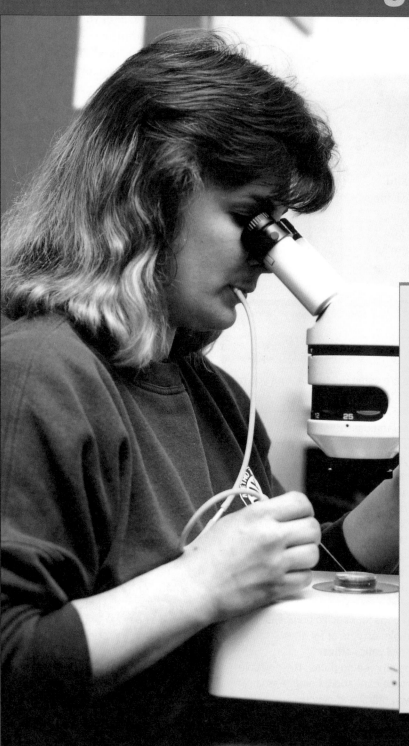

PERSONAL PROFILE

Career: Zoo biologist. "Over 15 000 species of mammals and birds are on the endangered list and will need our help to survive."

Interests: Crafts, including embroidery and cross-stitch; camping; scuba diving. "I'm currently president of our local diving club. I'm qualified as a divemaster so I can lead groups underwater."

Latest accomplishment: Obtaining a grant from the World Wildlife Fund to begin research on wood bison.

Why I do what I do: "I have a deep love and appreciation of nature, especially animals. Conservation of endangered species is part of my life."

I am: Independent, friendly, and confident. "I'm a kid at heart — I enjoy the simple things in life."

What I wanted to be when I was in school: "A veterinarian."

What a zoo biologist does

Biologists are scientists who study living things. Biologists may study whole ecosystems or just one kind of organism. They may work mainly in the field or they may do all their work in a laboratory. Karen Goodrowe's specialty is called reproductive physiology. "My job is to find ways to help captive animals reproduce successfully within a zoo environment," Karen explains. Many species are in danger of extinction in their natural habitats. "The world's zoos are working hard to ensure that species like the cheetah, wood bison, and elephant will still be with us in the future."

A zoo-ful of unusual problems

"I need to know a lot about how animals reproduce, and I need to keep up with new ideas in medical technology," Karen notes. "But often the problems I have to solve at the zoo aren't very technical!" For example, the usual way to find out if an animal is ready to breed or is already pregnant is to test its blood. But blood tests mean the animal has to be restrained and given a needle. "Take it from me — you can't easily get a blood sample from a gorilla on a daily basis," Karen warns. Like blood, urine can be analyzed chemically (in a test called an assay), so Karen and the zookeepers collect the gorillas' urine without disturbing the animals. They simply obtain it from the area within the enclosure where each gorilla tends to urinate.

Matchmaking

"The information we get from urine assays helps us plan the breeding program for the zoo's animals," Karen says. "For instance, we wanted a particular female zebra to have offspring. But other females would probably have produced young as well, if we had put the male zebra in with the entire herd. By using the urine assay, we could tell the keepers exactly when to put the male and this female together."

Help from technology

Karen and other biologists around the world have begun to collect living sperm and eggs from endangered animals. Sperm and eggs are reproductive cells. When a sperm and an egg unite in fertilization, the new individual that begins to grow has genes from each parent. Genes determine the characteristics of an individual. A gene may help an animal run faster, give it sharper eyesight, or make it more resistant to disease than some other members of its population. For populations to survive, a good mix of characteristics

is important. No one knows how an environment might change. "Thus, we don't know which characteristics might be most important to any particular species," says Karen. "That's why we need to try to collect sperm and eggs from as many different animals as possible. These sperm and egg cells can be chilled and kept alive for years in liquid nitrogen. Then, I use the technique of in vitro fertilization," Karen explains.

In vitro fertilization

In vitro fertilization is a technique in which an egg cell and a sperm cell are combined in a test tube or a dish. The embryo that forms is normal in every way except that it is not inside a mother's body. By means of surgery, the embryo can be placed in the body of a female (called a surrogate mother). There it will grow until birth. The embryo can also be frozen and stored for later development. "This means that we're providing for new offspring in the future, not just today."

Part of Karen's job is to use modern technology to help save animals like elephants that do not breed well in captivity, as well as those animals that are not breeding successfully in nature.

Many zoos are now refuges for the protection and breeding of endangered animals. Karen and her co-workers, for example, have helped this cheetah and many other animals to reproduce.

All in a day's work

Karen walks into her office around 8:00 a.m. each day. The zoo hasn't opened for visitors yet, but it's already busy. Karen's first task is to check her mail, which isn't quite what you'd expect. "Oh, I get some regular mail," Karen laughs. "But what I'm really after is news about the animals. This comes in two forms: electronic messages that I read using my computer, and daily report sheets from the zookeepers."

Mountain climbing — zoo style

On this particular day, Karen reads her mail, then pulls on her jacket and steel-toed boots. She and one of her research assistants will spend the next hour "mountain climbing." "The West Caucasian tur exhibit is built to match their natural habitat — rugged mountain country. It's not built for the convenience of researchers!"

Karen laughs. Today, Karen needs to collect urine samples from several female tur.

After climbing the turs' mountain to collect their urine, it's back to the laboratory with the collected samples. There, Karen and her assistant analyze the urine samples to determine which of the female tur are pregnant.

In the lab

Next, Karen begins work on an ongoing experiment. She wants to find out how to grow egg cells that are immature, in an incubator. At a certain stage, the egg cells can be fertilized by sperm. "This will let us save and use more egg cells from a female animal," Karen explains. "The death of a rare animal such as a snow leopard is a tragedy. If we can save her immature egg cells, as well as the few that are mature, we could possibly save up to 15 as-yet-unborn leopards."

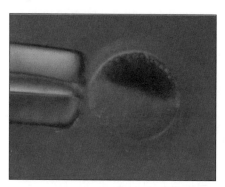

Magnification of an egg cell in a culture. The tip of a micropipette is shown on the left. The photograph on page 16 shows Karen using the micropipette to place the cell in the culture.

The daily report sheets tell Karen about the health of the zoo animals and about the arrival or departure of animals.

Tur are from natural habitats in Europe and live high up in mountains, just below the snow line.

Finding funds

Later, Karen works at her computer, writing a grant proposal. It's an important part of her work. "I think about what type of research or experiment we need to try next. Then I write a proposal that explains my ideas and the amount of money or equipment I will need. Fortunately, there are groups and organizations that want to help endangered species by financing scientific research like ours." It's something Karen really enjoys. "It's fun to put your ideas on paper and find that someone else believes in an idea enough to put money toward its success."

Taking care

Animal care is Karen's next task. She's the chairperson of the zoo's internal Animal Welfare and Research Committee. This committee makes sure that each animal's well-being comes first. It looks at research proposals that might involve the zoo's animals. The committee members make sure the animals will not be made uncomfortable in any way during the research.

End of the day

On her way out at the end of the day, Karen stops by the gorilla enclosure. "I love them all, but Josephine is my favorite," she confesses. "She's so intelligent." Karen describes one of Josephine's many accomplishments: "When I stand by her cage, Josephine comes up to me with a small wood chip in her hand. It's a game we play. I trade her a peanut for the chip. For a moment, I forget about how threatened her species is. I just enjoy the fact that I can be so close to such a wonderful creature."

It's a Fact

"Most people are surprised to learn that of all the species of cats in the world, only one — the domestic cat — is not threatened or endangered," Karen remarks.

Activity

Making your own observations

In order to help endangered species reproduce in captivity, biologists like Karen need to know as much as possible about how animals reproduce. They do this by making close observations of the animals' behavior. Here are some ways you can make your own observations.

- Raise tropical fish, such as guppies or mollies, that reproduce readily in small tanks. Find out the conditions that the fish and their young need for survival.
- Watch for birds gathering nest material in the spring. Continue to observe any birds that are actively nesting. Record your observations over the entire breeding season.

Alvaro Jaramillo, shown above, is a field biologist who studies animals in their natural habitats. Through field observations such as those of Alvaro and other biologists, we are learning more about the lives of animals like the slaty-tailed trogan, shown here in a tropical rain forest, its natural habitat.

How to become a zoo biologist

Before specializing in a particular field of research, a biologist takes a college or university program focusing on science. Karen's career choice sprang from her ambition to do what she enjoyed. She had always loved science. "I could take or leave math," Karen recalls, "but I kept taking it nonetheless!" She studied biology as an undergraduate, because she planned to become a veterinarian.

For her senior project, Karen researched the effect of DDT on animals living in wildlife refuges. This sparked her interest in conservation. "I realized that conservation meant that some animals were going to need extra help to survive. Just giving them space wasn't enough." So Karen applied for graduate studies in biology that involved veterinary science and conservation. There were about five people in the world working with the reproduction of zoo animals when she started out. "I was setting my own course, though I didn't know it at the time. I knew what interested me, and I looked around for a way to do it."

The way opened up for Karen in graduate school. "It was great," Karen remembers fondly. "Lots of hands-on work with animals. I loved it." She then wrote to all the major zoos in North America. She asked for either a job or a chance to continue her studies with them. "They all said no. I had to have a Ph.D. degree first. Then I contacted Dr. David Wildt." Dr. Wildt was starting a new program at the National Zoo in Washington, D.C. My interests matched Dr. Wildt's new 'cat project' perfectly," Karen smiles. "It turned out I was the first-ever zoo-oriented student." This wasn't Karen's only first. Dr. Wildt's "cat project" produced the world's first test-tube kittens!

Karen then obtained her Ph.D. degree. She was hired by the Metro Toronto Zoo to set up its reproductive physiology research program. "There are now many jobs in zoo biology that don't require a Ph.D.," says Karen. "An undergraduate degree is enough for many of them."

It's a Fact

On a worldwide basis, it is estimated that one species per day is becoming extinct.

*I*s this career for you?

To work with endangered animals takes dedication. "You need to deeply care about conservation to keep going," Karen says simply. "Sometimes the meetings seem endless, or you're kept from your research by other tasks. It can be hard to stay focused. But it's all part of the real job, saving species from extinction."

Determination and organization

Being dedicated is important, but you also need a very determined attitude. "If someone says you can't do something, just try from another direction," Karen says. "You usually find out you can do it, after all."

An important part of scientific research is to make careful records about experiments or about an animal's medical history. "Someone else might need your records years from now," Karen explains. "As well as record-keeping, you have to deal with information coming at you from all directions. So being well-organized and having good powers of concentration help."

Working with others

Karen pauses, then says, "One of the most useful skills you can have is the ability to get along with other people.

Karen with the world's first test-tube kittens.

That's really important in scientific research, believe me. No one works alone. I interact with many different kinds of people all the time. These people include veterinarians, zookeepers, research assistants, the visiting public, and others."

Career planning

If you live near a zoo, ask permission to "job shadow" a zookeeper or zoo biologist: take notes as you watch the person "on the job."

Making Career Connections

Most zoos have "adopt-an-animal" programs. Organize a fundraising event so you and your family or friends can help support an endangered animal.

What interests you most — birds, snakes, fish, monkeys, whales, or deep-sea squid? When you're asked to do a class project, write about the animal species that most interests you. Include information on its current status.

Find out what careers are possible in biology by writing to a nearby college or university. A guidance counselor or a librarian can also give you ideas on career exploration.

Getting started

Interested in being a zoo biologist? Here's what you can do now.

1. Plan on taking courses in science, especially biology, in high school. Don't forget math and chemistry.
2. Learn about conservation and pay attention to issues related to animals. Read magazines such as *International Wildlife* to learn about the problems animals face.
3. Join a group that works with animals, such as a 4H club. Or help at any job that involves animals, such as in a dog kennel or a veterinary clinic.
4. Keep domestic pets and learn about their needs.
5. Many organizations, including libraries and museums, invite conservationists to be guest speakers. Check in your newspaper for announcements of such talks. Don't hesitate to introduce yourself to the speaker. People who work in conservation are usually interested in involving others.

Related careers

Here are some related careers you may want to check out.

Curator
Oversees the kinds and numbers of animals that a zoo has. For example, if the zoo needs another animal of a certain kind, the curator locates and acquires it. (This is usually done through "breeding loans" with other zoos.) Exchanges information with curators at other zoos. Designs new exhibits.

Zookeeper
Entrusted with the daily care of animals. Notices changes in an animal's behavior and notifies the veterinarian. Keeps animals active and healthy.

Zoo veterinarian
Treats injuries or diseases afflicting zoo animals. Advises keepers about the animals' diet and care. Assists when animals are giving birth.

Field biologist
Studies living things in their natural environment. May specialize in one

particular organism or one kind of environment, such as desert or rain forest.

Research assistant/technician
Carries out day-to-day tasks to complete research projects.

Future watch

The career prospects for zoo biologists are very good. Research into the reproduction of wild animals is expanding. "So many species are already endangered, and more become endangered each day," Karen comments. "Sadly, this means there'll be even more need for human help in the future."

Derek Choong

Veterinarian

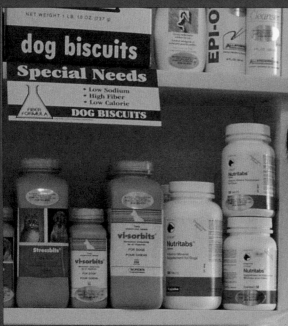

PERSONAL PROFILE

Career: Small animal veterinarian. "A lot of what I do prevents health problems from happening to pets."

Interests: Photography, downhill skiing, travel. "Travel was something I just dreamed of doing when I was a student. Now I'm able to take some of the trips I couldn't afford before."

Latest accomplishment: Opening a small animal clinic of his own, in partnership with another veterinarian.

Why I do what I do: "I have a strong sense of caring and respect for animals."

I am: Determined, goal-oriented, friendly, and compassionate. "You couldn't survive as a small animal vet without being good with people."

What I wanted to be when I was in school: "I was torn among three interests: veterinary medicine, computer science, and human medicine."

What a veterinarian does

A veterinarian takes care of animal patients the same way a physician takes care of human patients. "For example," Derek explains, "I do annual checkups to make sure healthy pets stay that way." Part of a checkup includes giving vaccinations to prevent diseases such as rabies. "Nothing is worse than seeing an animal die from a disease that could have been prevented," Derek says, shaking his head.

Health checks

"Just as a family doctor might need a test done on a blood sample to check on your health," Derek continues, "I depend on similar tests to assess the health of the animals I see." Although Derek or his partner do most tests at the clinic, a few are done by veterinary technicians in other laboratories. "We can do urine tests here, for example, but we send blood samples to a laboratory."

Derek checks the animal's heart and lungs using a stethoscope, then feels for any unusual lumps or swellings. "A healthy animal has a good coat, clear and alert eyes, and clean ears," Derek says. If necessary, Derek cleans the animal's teeth and clips its nails. "Many people don't realize that dogs and cats can suffer from painful tooth decay. Long nails can also lead to problems."

As a routine part of the examination, Derek checks each animal's teeth. "People go to a dentist for their teeth, to an optometrist for their eyes, and to other specialists depending on their needs," Derek says with a laugh. "To my patients, I'm all those things in one!"

Out of the ordinary

A small animal clinic like Derek's often has pocket pets as patients. "Pocket pet" is a name given to any small animal, such as a gerbil, hamster, or rabbit. "Big or small, we keep careful, detailed records of every animal that comes to the clinic. These records help us follow the animal's medical history, and show what care the animal received."

When something is wrong

A large part of what a veterinarian does is like solving a puzzle. "First, a pet's owner notices some change in the animal's normal behavior and describes this change to me," Derek says. "Sometimes the animal has diarrhea or is vomiting. At other times, it's a less obvious change, such as the animal no longer wanting to play.

"Next, I combine the owner's observations with the animal's history, and with the results of a physical examination of the animal. Since our patients can't tell us what hurts, it is important to be able to assemble these pieces of the puzzle."

But we didn't want kittens

"Each year, Humane Societies receive hundreds of thousands of unwanted puppies and kittens," Derek's tone is frustrated. "Homes cannot be found for most of these animals.

"Most people who keep pets don't plan on their animals having babies," Derek explains. "I advise these owners to bring in their puppy or kitten at about six months of age for the surgery that prevents reproduction." This surgery is of two kinds: neutering for male dogs and cats, and spaying for female dogs and cats. Neutering is the removal of the testes. Spaying is the removal of the ovaries and the uterus. Both operations are straightforward and pets go home the next day. "If everyone with a pet acts responsibly," Derek comments, "pet populations can be controlled."

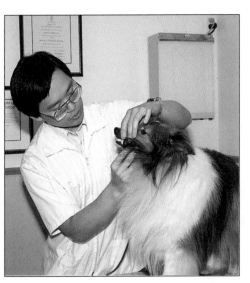

"Rabies only seems rare," Derek warns. "If pets weren't vaccinated, the disease would spread from infected wildlife to pets and then to humans."

All in a day's work

Derek unlocks the clinic doors just before 8:00 a.m. each morning. After turning on the lights, he unlocks several drawers. "As in any medical facility, we keep chemicals locked up when not in use," Derek notes. "Also, our records on each patient are as confidential as your physician's records are about you."

Next comes the important task of checking the hospital list. "This is an up-to-date list of all the animals currently hospitalized in the clinic. These animals are recovering from surgery or are being treated." During the next hour or so, Derek goes from patient to patient, making sure they are doing well, and changing bandages if necessary. "My assistants, my partner, and I talk about each animal's progress. Anything that one of us happens to observe could help us look after the animal better."

Into surgery

From 8:30 a.m. until 10:30 a.m., Derek does surgery. "We schedule surgery for the morning so the animals that were anesthetized for their operations will wake up during the afternoon, while the veterinarians are still at the clinic," Derek explains. "Occasionally an animal is slow to wake up, or seems uncomfortable. I want to be there to help." In some cases, the animals will be ready to go home before the end of the day.

While Derek reviews each animal's medical record, his assistant prepares the surgery room. "Cleanliness is the single most important way to prevent infection. Everything from the floor up is scrubbed with disinfectant," Derek remarks. "After I've washed, I put on a mask and gown."

First, the animal is given a drug to relax it. After 10 minutes, the main anesthetic is given, which puts the animal into a deep sleep. The area of skin to be operated on is shaved and cleaned. One of the assistants or the other veterinarian will assist Derek during the operation.

"Doing the actual surgery is usually the shortest part," Derek says. The cut, or incision, into the skin and muscle is kept as small as possible. "Because I've done this kind of surgery so often, both in school and in the clinic," Derek explains, "I can do the surgery neatly and quickly. This makes it easier for the incision to heal." The animal is then moved to a quiet place to recover.

After surgery, Derek and his partner spend at least an hour adding details to each patient's records. "The medical record is vital. I write down what happened during the surgery, how much anesthetic was used, and anything I observed about the animal."

Checking the eyes, ears, and teeth is an important part of every examination. "Most animals let me handle them without a fuss," Derek comments. "They're a bit nervous about being up on the table, but they respond well to a calm voice."

Unwelcome guests

Warm weather may bring some unwelcome guests — fleas — to feed on the blood of your pet. Fleas are agile and quick, leaping farther (for their size) in a single bound than any other living thing! Because of this jumping ability, fleas lurking in the grass can hitch a ride on a passing dog or cat. Once there, they wriggle their way to the animal's skin, especially behind the ears and on the neck. The pain of flea bites makes an animal's life miserable. Worse, fleas can bite humans as well and they sometimes carry disease. "To get rid of fleas," Derek recommends, "you have to treat the home as well as the animal." This is because the female flea lays her eggs on the ground, such as in the fibers of a rug. When old enough for a blood diet, her offspring will leap onto the next warm-blooded animal to walk by, starting the infestation all over again.

Newcomers to the clinic complete a form that begins the pet's medical record.

Animal control to the rescue

All Derek's appointments have to wait if there's an emergency. One day recently, an emergency patient wrapped in a blanket arrived in the arms of an animal control officer. "A nice, short-haired pointer was struck by a car and left on the road," Derek recalls. The animal control officer had brought the dog to Derek as quickly as possible. It was badly cut and bruised.

"The dog was conscious, but seemed dazed. I was sure it was in shock." An animal in shock is not receiving enough blood to its body parts and may die. Derek immediately took the dog's temperature and blood pressure. Both measurements were far

Some people like to keep unusual pets, such as this sulfur-crested cockatoo. Veterinarians need to know how to treat any animal that is brought to them. "The conditions that cockatoos need to reproduce successfully are now known," says Derek. "This should help reduce the illegal sale of wild birds."

too low, confirming his diagnosis of shock. Right away, he made the dog comfortable and began giving shock treatment by administering fluids and medication into the animal's bloodstream.

When its temperature and blood pressure show that the animal has survived shock, Derek will then operate on its injuries. "We were able to operate later that day," Derek recalls. "The dog's owners were contacted by Animal Control, and the

dog was soon well enough to go home. It's good to be part of a happy ending like that."

The end of the day

Before locking up for the night, Derek makes sure all the animals in the hospital room are comfortable and have received any medicine they require. The surgery is clean and ready for tomorrow's operations.

Derek locks the drawers, turns out the lights, then listens for a moment to be sure everything is peaceful. "It's a satisfying moment," he smiles. "Not many jobs give you that feeling at the end of each day."

Pet	Wild animal with similar habits
dwarf rabbit	pika
domestic dog	timber wolf
domestic cat	bobcat or ocelot
goldfish	carp
gerbil	kangaroo rat
horse	zebra
ferret	weasel or mink
domestic rat	Norway or brown rat
canary	goldfinch

Activity

The wild side

An adaptation is a behavior or body part that helps a species survive in its usual environment. For example, being able to sneak and pounce helps a cat capture other animals for food. How can you find out more about the adaptations of a pet animal? One way is to compare a pet you know to a similar wild animal. Use the information in the chart on the right to help find a good comparison.

Look in reference books for information on the adaptations of the pet's wild "cousin." Try to find answers to these questions: What kind of food does the wild animal need? How does it get its food? What is the relationship between this animal and other organisms in its environment?

How to become a veterinarian

"This is a career you pick because you're interested in animals," Derek remarks. While growing up, Derek worked with his father, who was a large animal vet. He knew from this experience that he wanted to be a veterinarian. But wanting isn't enough. Every year, more students apply for veterinary medicine than can be accepted. "You have to do well in high school," Derek points out, "especially in science and math."

After high school graduation, Derek began a university science program. "This is the critical time, without a doubt," Derek observes. "Depending on where you are, you can apply to veterinary college during or after your science degree. High achievement is important."

Veterinary college... and after

Derek was accepted into a college of veterinary medicine the first time he applied. But many of his science classmates were not accepted right away. "A lot of students reapply several times. Persistence helps!"

The first part of the veterinary program is called the "pre-vet" year. Students who successfully pass the pre-vet courses can take the remaining four years of veterinary medicine.

After the third year, students work in clinics for the summer. This is just like an internship in human medicine. Then the fourth and final year is spent entirely in clinic work, applying knowledge to actual cases. After graduating from college, Derek and his classmates worked as associates with established veterinarians. "Internship gave us a chance to improve our skills," Derek comments. "It also gave us a chance to earn some money and pay back student loans. The cost of so many years of education is high, but it's worth it in the long run."

Is this career for you?

Do you like working with animals? Do you enjoy learning new information and using it in a wide variety of situations? If so, veterinary medicine could be a good career for you. You'll also need good analytical skills. "You have to keep on top of things even after graduation," Derek reports, "because technology in veterinary medicine is constantly advancing and new medicines are being discovered all the time."

If you're thinking about being a vet, Derek advises, "You need to know how to make people comfortable and at the same time ask them the right questions." Two other characteristics are essential for any veterinarian. "You need to be good with your hands," Derek observes. "And you'd better not be squeamish."

There is only one thing Derek doesn't like about his work. "The long hours," he says without hesitation. "The clinic must be open at hours convenient to the public, but we also have to be here for the hospitalized animals." How does Derek cope? "My partner and I work in shifts. And we depend on our assistants. Sharing the workload helps cut down the number of hours worked."

Many veterinarians, such as Dr. Jacqueline Starink, specialize in the care of large animals. Often, these veterinarians travel to their patients, rather than operating a clinic like Derek's.

Career planning

Ask a local veterinarian if you can watch him or her for a day or a half-day. If the veterinarian agrees, record the tasks done so you are prepared to describe the work to your family or a friend.

Making Career Connections

Find out about work programs through your school. If there isn't one, approach your local veterinarian. Ask for hands-on summer work as a kennel attendant or part-time receptionist. This is the single best thing you can do.

Do research projects on living things that especially interest you. Your local library and government wildlife agencies are good sources of information.

Write to a veterinary association to obtain information about becoming a veterinarian. Your school guidance counselor or a librarian will be able to help you.

Getting started

Interested in being a veterinarian? Here's what you can do now.
1. Volunteer to help at a local veterinary clinic on weekends. This work will help you see firsthand the different tasks performed in a veterinary clinic.
2. Provide a pet-sitting service in the summer months.
3. Get a summer job on a farm.
4. Attend open houses at veterinary colleges and talk to the staff there.
5. Keep pets and learn about their needs.
6. If you have trouble with a subject, now's the time to ask for extra help. You need to do well to be accepted by a veterinary college. Take courses in science, especially biology, in high school. Don't forget math and computers!

Related careers

Here are some related careers you may want to check out.
Veterinary technician
Assists in a veterinary clinic.
Pet groomer
Cares for the appearance of dogs, cats, and other animals. Is usually self-employed, working under contract to several veterinarians.
Animal control officer
Assists stray or injured animals. Enforces local animal control bylaws. Traps nuisance wild animals and takes them where they can be safely released. Responds to complaints and emergencies involving animals.
Scientist
Researches animal care for universities, governments, and private industries.

Wildlife rehabilitator
Works in a clinic that receives injured wildlife. Treats animals so they can be returned to the wild. May also be a fundraiser and coordinator of volunteers.

Future watch

The career prospects for veterinarians are good. Pets are important in people's lives and good pet care includes regular visits to the veterinarian. Also, agriculture is depending more and more on veterinarians in order to use new technology to breed better livestock animals.

Bonnie Powell

Veterinary Technician

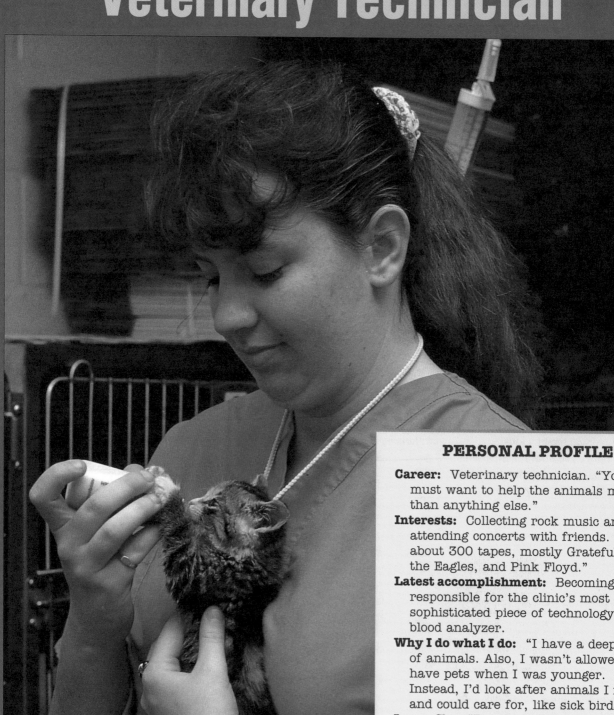

PERSONAL PROFILE

Career: Veterinary technician. "You must want to help the animals more than anything else."

Interests: Collecting rock music and attending concerts with friends. "I have about 300 tapes, mostly Grateful Dead, the Eagles, and Pink Floyd."

Latest accomplishment: Becoming responsible for the clinic's most sophisticated piece of technology, the blood analyzer.

Why I do what I do: "I have a deep love of animals. Also, I wasn't allowed to have pets when I was younger. Instead, I'd look after animals I found and could care for, like sick birds."

I am: Shy. "But this job has made me learn to be outgoing with people."

What I wanted to be when I was in school: "A veterinarian, no question."

What a veterinary technician does

In many ways, a veterinary technician's work is like the work of a nurse in a physician's office or at a hospital. Bonnie sums it up: "A veterinary technician is trained to help the veterinarian and to provide care to the animals."

What many people don't realize is that veterinary technicians have special skills. "In many cases, a technician is more knowledgeable about tests and equipment than the veterinarian is," Bonnie notes. "That makes sense, because it means the veterinarian can concentrate on the animals and their health, while technicians keep up-to-date on test procedures in the laboratory."

While many veterinary technicians work in typical veterinary clinics, Bonnie has chosen to work in a rather unusual place. "There was a need for veterinary care to be available any time of day — not just during office hours," Bonnie explains. "So several veterinarians pooled their resources and built an after-hours emergency clinic. We're open all night and on weekends. The veterinarians take turns coming in. But there are always veterinary technicians on staff full time."

Like the emergency room in a hospital, Bonnie's clinic may have as few as three patients in a night, or as many as 30. "Weekends are busier," she points out.

Laboratory work

When you go to your family physician, you might be asked for a sample of your blood or urine. These samples are analyzed, or tested, in a laboratory. The results of these tests give your physician a better idea of what is going on inside your body. "At the emergency clinic," Bonnie says, "We do all our own tests. There's no time to send samples to another laboratory. So analysis is a very important part of my work."

An animal's urine analysis provides many useful clues to its health. The chemicals in urine can be tested to check on the animal's kidneys and digestive system. "We look at a sample of urine under the microscope," Bonnie explains. "If I see small crystals in the urine of a neutered male cat, the animal could have feline urological syndrome. This is a serious disorder that needs immediate treatment, before the crystals build up in the animal's bladder."

The microscope is only one tool. Bonnie's pride and joy is the blood analyzer. "I put a drop of blood on a clean slide and place it in the analyzer.

It then prints out information about the blood." The machine itself requires a lot of care. "One thing I do routinely is check the calibrations, or settings, on the analyzer to make sure it is measuring precisely."

Kinds of care

While the animal is being cared for at the clinic, Bonnie changes its bandages and checks its condition regularly. She takes each animal's "TPR," which is an abbreviation for temperature, pulse (heart rate), and respiration (breathing). "We have a large board on which the veterinarian writes down the care every patient needs," Bonnie says.

"Animals need emotional care, too," Bonnie points out. "We can do a lot to comfort and relax them. Sometimes a hot water bottle to cuddle with helps, or just a friendly voice. Even a few minutes stroking and patting can work wonders."

Lastly, Bonnie does a lot of client education. "I explain to the pet's owner how to give medication, how to change bandages, and what kinds of food or special care to give the pet."

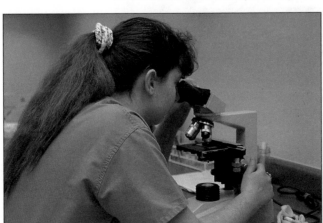

Bonnie checks a urine sample for crystals.

Bonnie places a sample of the patient's blood in the analyzer.

All in a night's work

When Bonnie arrives for work at midnight, she changes into her scrub suit and puts on a laboratory coat. She then spends time checking the cases (patients) already at the clinic. One patient, a cat with pneumonia, is in the incubator, with an oxygen supply to help it breathe. The veterinarian has asked for X rays of the cat's lungs and for an IV (intravenous needle) with fluids and antibiotics to be checked. Before she starts, Bonnie reaches into the incubator and strokes the animal gently for a few minutes.

There are no schedules or appointments at the clinic. "If a lot of animals arrive at once, the vet will put each patient into a cage, write down what tests are needed, and go on to the next one." Bonnie performs the tests and brings the results to the veterinarian.

When Bonnie isn't busy with animals, she spends her time doing routine tasks. "I put fresh supplies on the shelves. I then check the equipment and run standards on the analyzer."

Drop everything!

One night a big golden retriever was brought in. The dog had been giving birth to puppies when the owners noticed that her labor had stopped after one puppy was born. It was plain to see that several more puppies were still to be born.

"The vet decided the dog needed help to go into labor again. We gave the dog a drug to restart her contractions. I then X rayed her. The X ray showed a mass of tiny spines and skulls, the remaining puppies.

"We all helped the mother, encouraging her and taking each puppy as it was born. She had five more." Bonnie's task was to break the birth sac around each puppy, then pass the pup gently to the other technician, who would make sure it was breathing properly. The vet cut the umbilical cord and passed each puppy back to the mother.

"What a time we had," Bonnie remembers with a smile. Four hours later, the mother dog and the pups were able to go home.

Another emergency

A dog that had caught its hip under a fence was the next emergency. "I took X rays, prepped the dog for surgery, and monitored its heartbeat while the vet prepared to operate. As you can see, the patient came through just fine."

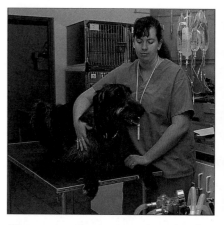

After surgery, the dog is awake and well on its way to recovery.

The next case arrived half an hour later. "A man brought in his cat and four three-week-old kittens. He described the mother cat as listless. She was trembling." The veterinarian asked Bonnie to test the cat's blood for calcium levels. "We knew it was probably a problem with the cat's diet." The cat was *not* cooperative. "It took two of us to get a blood sample — one holding the cat wrapped in a towel!"

The blood analyzer gave the expected answer. The cat was suffering from a lack of calcium. Nursing animals need more calcium than usual, because they are producing calcium-rich milk. While the mother cat stayed at the clinic that night, receiving calcium through an IV, the owners had to feed the kittens using nursing bottles — every four hours.

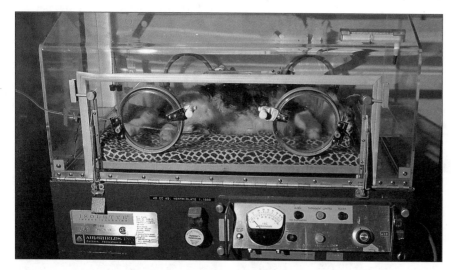

This cat is kept warm and is supplied with oxygen. "We don't know if she'll recover yet," Bonnie says softly. "But she's in the right place."

Near morning

It's now 3:00 a.m. "Usually it calms down by this time, although we sometimes get night finds." A "night find" is an animal that would normally be picked up by an animal control driver and taken to the pound. But animal control staff only work days. "We once treated a deer that had been hit by a car. We all think of deer as gentle, quiet creatures, but try to X ray one!" laughs Bonnie. "I still have the bruise where it kicked me."

Bonnie spends the rest of the night checking patients and catching up on paperwork. "We sit down for dinner at about 5:30 a.m.," she says. At about 6:00 a.m., Bonnie checks all the animals in the clinic and writes a report on each one. A copy of the report will go with the animal when it leaves.

"Every animal is picked up by its owner (or animal control) between 7:00 a.m. and 8:00 a.m.," Bonnie explains. "When the clinic closes at 8:00 a.m., we can relax a bit." Bonnie feeds the clinic's three pet cats, does some accounting, orders supplies from veterinary supply companies, and sterilizes all the surgical instruments. Finally, at about 10:00 a.m., she's ready to go home.

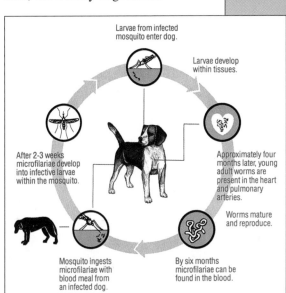

Larvae from infected mosquito enter dog.

Larvae develop within tissues.

After 2-3 weeks microfilariae develop into infective larvae within the mosquito.

Approximately four months later, young adult worms are present in the heart and pulmonary arteries.

Worms mature and reproduce.

Mosquito ingests microfilariae with blood meal from an infected dog.

By six months microfilariae can be found in the blood.

Heartworm cycle

"Heartworm is a parasite that's becoming more common in dogs. It is a kind of roundworm that lives in the dog's heart and blood vessels," explains Bonnie. Heartworm is spread from dog to dog by mosquitoes. The worms block the flow of blood through the heart, and an infested dog may be killed or permanently disabled. The only way to detect heartworm is to examine a blood sample. Fortunately, dogs can be protected if they are given heartworm medicine during mosquito season. The medicine kills any worms that enter the dog's blood.

Activity

Life-beats

Subject	Resting heart rate (beats per minute)
adult dog (medium-sized)	100
puppy	130
adult cat	120
kitten	140

Compare your pet's heart rate to those listed in the chart. Place your hand between the animal's front legs and its chest. Feel the chest for the heartbeat. (Or, place your ear against the animal's chest and listen for its heartbeat.) If your pet's heart rate differs from the rate in the chart, don't be concerned. Many factors affect heart rate, and normal healthy animals can be very different from each other. For example, small breeds of dogs normally have more rapid heart rates than larger breeds. What might account for this and other differences? Think about how your own heart rate changes when you are anxious or involved in exercise. Your heart has to beat more quickly and strongly to supply all your active body parts with enough blood. Heart rate is also affected by body mass, and by how much of that mass is fat. Ask a veterinarian about how being overweight affects the heart rate and the health of pets.

Challenge

If you have a pet goldfish, try this. When the fish is resting, count the number of times its gill flaps open and close in 15 seconds. Then multiply that number by 4 to obtain the breathing rate per minute. At the time you normally feed your fish, observe how the fish behaves. At this time, take its breathing rate again. What do you notice? How long does its breathing rate take to return to the resting rate?

How to become a veterinary technician

"I wanted to be a veterinarian," Bonnie remembers. "But I felt I could not do well enough in courses to become one. I had worked part time in a veterinary clinic, and knew the technicians had just as much to do with the animals as the vets did. So that's the direction I chose."

High school years

Bonnie concentrated on doing her best in biology, chemistry, and math. "They were the major focus. But I did a lot of extracurricular things, too." When she was 14, Bonnie was a volunteer during the summer at the local Humane Society. "By September, they offered me a part-time paying job."

The next year, Bonnie got a job as receptionist at the emergency clinic. I worked during the day on Saturdays and Sundays." After a while, an opening as a kennel attendant became available at the clinic, and Bonnie took it. "My experience as a receptionist helped. I'd learned not to be shy. You have to see yourself as part of a team."

The technician program

After high school, Bonnie applied to the veterinary technician program. "This college program takes two years. They only accept a few new students a year, so it's important to keep your performance as high as possible." During the first year of the program, Bonnie studied physiology, biochemistry, and immunology. "We learned basic animal handling techniques, such as applying bandages." The college provided animals for the students to handle. Bonnie smiles. "We adopted them at the end of the year."

The second year of the course is more specialized. Bonnie learned about surgery, radiography (X rays), and animal nursing. "There is a lot of hands-on work." The courses range from clinical chemistry to hospital management.

Valuable connections

In most veterinary clinics, at least some of the people you meet will be high school students working as volunteers or assistants. Karen Machin had many such jobs when she was a student. She also worked as a stable groom, and as a camel walker at a local zoo. Now Karen is a large animal veterinarian. She is working with other veterinarians and technicians to try to prevent the spread of rabies in wild animal populations in Africa. Karen urges students to look for volunteer and part-time work in high school and during the summers. She says this type of work gave her valuable experience for her job today.

Is this career for you?

"Veterinary technicians are taught to have dispassionate compassion," Bonnie says seriously. "That means we care about the animals, but we don't let ourselves get wrapped up emotionally." This is vital in any medical profession. "Because I'm dispassionate, I can give a needle to a kitten, knowing the medicine I'm injecting will help it. Because I care, I can make myself cleanse a badly infected wound. How else can these animals be saved?"

It certainly helps to be quick and accurate under pressure. If you are the kind of person who pays attention to details, even in a rush, this is the job for you. There's no room for mistakes in the lab work — a botched test could mean the animal will die.

"You'd also better enjoy working with people. Usually at least four different people are working with every animal. This is no job for someone who is hard to get along with."

Variety and excitement

Bonnie likes the variety of cases that come to the emergency clinic. "It's an exciting place to work. There's no set routine to bore you," Bonnie smiles. "And I like working with great people who know their jobs. We all care deeply for animals. There's an attitude of valuing each other's skills. I'm glad to come to work. Even at midnight!"

Career planning

Making Career Connections

Do you have a dog or a cat as a pet? If so, ask a veterinarian to show you how to clip its nails or check its teeth. Then be responsible for carrying out these tasks.

Ask a veterinary technician if you can watch him or her at work for part of the day. Follow up your job shadowing (as this career investigation technique is called), by keeping a written log of the technician's activities.

Think of the living things that interest you, and do school projects to learn more and more about them.

Write for information on veterinary technician courses offered in your area. Ask your school guidance counselor or a librarian to help you find a list of appropriate schools to contact.

Getting started

Interested in being a veterinary technician? Here's what you can do now.

1. Take math and science in high school. You'll use what you learn every day on the job. Doing well in these subjects is especially valuable when you apply to college.
2. Volunteer to help at your local veterinary clinic, kennel, pet shop, or Humane Society.
3. Look for summer jobs that let you work with animals. Many clinics, for example, hire summer students as kennel attendants.
4. Keep pets. Learn how to handle them gently but with control.

Related careers

Here are some related careers you may want to check out.

Wildlife rehabilitation assistant
Helps injured or sick wild animals to recover so they can be released back to the wild.

Animal control driver
Often the first person to encounter an injured animal. Also responds to citizens' calls about animal pests. Uses humane methods to capture wild animals that have become a nuisance to people, releasing the animals elsewhere.

Entrepreneur
Establishes own business, such as pet-sitting service, grooming service, or kennel design company.

Future watch

When the economy is bad, people can't afford as much money for pet care. So veterinarians have less money to hire technicians. But there's good news. More and more emergency clinics are opening up. And people are keeping more pets than ever. Overall, the future for this career looks good.

Plant Scientist

PERSONAL PROFILE

Career: Plant scientist. "I work with plant cells to produce improved plants and crops."

Interests: Friends. "We always make time for each other — even if it's just a break from the lab to sit in the lounge and talk."

Latest accomplishment: Having his work published by an international journal of applied science. "The editors of the journal choose research they feel is particularly valuable to other scientists. To have your work selected is quite an honor."

Why I do what I do: "I want to make a difference. I believe that my work will help solve the food crisis that exists in many parts of the world."

I am: Shy, but friendly. "I need some time with a new person or situation before I can relax and be myself."

What I wanted to be when I was in school: "An airline pilot."

What a plant scientist does

Some plant scientists work mainly outdoors, studying the ecology of different plants. Others, such as John Afele, work in laboratories. John Afele is a plant biotechnologist. "Biotechnology is the use of living things or extracts from living things to produce a certain product," John explains. "Biotechnology helps produce more offspring from especially useful parent plants."

Custom crops

Plant scientists are always looking for individual plants with useful characteristics. "Imagine you have a garden full of tomato plants," John explains. "One plant produces twice as many tomatoes as any of the other plants. If you collect and grow seeds from that tomato plant, you might immediately get plants with the "more-fruit" characteristic — or you might not. Using biotechnology, we can one day produce as many identical copies of that plant as we wish."

John is talking about a complex technique called cloning. In cloning, exact copies of an organism are grown from single cells.

Plant-tech

Cells inside a living organism receive all the nutrients and conditions they need to stay alive. To grow cells *outside* the organism, in a tissue culture system, John needs to duplicate those conditions in the laboratory. The cells he cultures come from seeds. "A seed contains the beginning of a new plant, called the embryo. The cells of the embryo are ideal for culturing. They contain the genetic information needed to grow the plant." John works under a microscope to find and handle the cells. "I place them in a culture dish, then wait for things to happen."

If all goes well, in a couple of weeks a lump can be seen growing in the culture dish. This lump is called a callus. John keeps the dishes with the growing cells in a special cabinet with controlled temperature, light, and humidity.

Artificial seeds

The cells growing in the culture will not sprout roots and leaves by themselves. John must "convince" some of the cells to form a seed. "I place chemicals on the cells that the plant normally uses to control its own growth," John says. "This is the trickiest part of my work. I try to mimic exactly what happens in a complex living organism, using only a test tube and a mixture of chemicals."

Each of the artificial seeds John produces in this way is a tiny embryo. Embryos grow immediately and rapidly. "This is a problem," John points out. "We need to be able to store the seeds, sometimes for several years." To overcome the tendency of his artificial seeds to start growing, John dries them. "The dried artificial seeds can be stored indefinitely, safe from disease or fungus. We simply plant them when we need to."

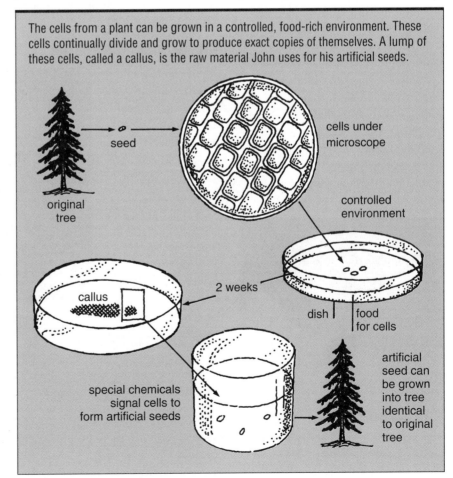

The cells from a plant can be grown in a controlled, food-rich environment. These cells continually divide and grow to produce exact copies of themselves. A lump of these cells, called a callus, is the raw material John uses for his artificial seeds.

All in a day's work

John usually works on several different projects at once, or else on one project that has several different stages. This calls for careful organization. "I plan my experiments so I need to pay attention to only one experiment at a time. Living cells can't wait while I do something else."

Picky eaters

On this particular day, John is getting dishes ready for his next set of plant cells. All dishes must be filled with the right media, the mixtures that will hold the cells and provide them with nourishment. John spends the morning preparing the proper "diet" for the cells. "The dishes contain everything the cells need: sugar, vitamins, water, and so on. Amounts matter, too. Cells need to have exactly the right amount of each vitamin, mineral, or other substance."

John uses his own experience, but he relies on the work done by his peers as well. "You don't work alone in science. I read scientific journals from the library to find out how other scientists have tried to solve similar problems. I try their ideas as well as my own."

John says, "As I work, I am always asking myself questions like "why?" and "what if?"

Battling bacteria

John wears gloves as he works and the laboratory is kept sparkling clean. "The media I use to grow plant cells are also attractive to bacteria," he says. "And bacteria grow much more quickly than cells. If bacteria get into the culture dish, a couple of weeks later I'll have a bacteria culture — not a plant one!" There is another concern. "I want to produce disease-free seeds so the plants that grow from them are also healthy. Thus, I work under sterile conditions, using gloves, and I sterilize anything that touches the cells. By doing this, I can protect the plants from disease-causing bacteria, fungi, and viruses. Yes, in case you're wondering, plants can get diseases, too!"

Microscopic gardening

Once John has a lump of cells growing well in his dish, he must take regular care of the culture. "Just as a house plant can outgrow its pot, my

These lumps of cells don't look much like a spruce tree, but each microscopic cell of each callus is capable of growing into a tree.

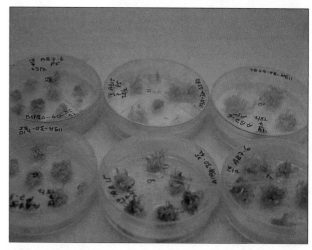

cultures can do the same." He carefully trims the excess growth and puts the culture into a new dish. "Cared for properly, a callus will continue to produce new cells for years." John has been snipping material for his artificial seeds from one blue spruce callus for several months.

These tiny spruce trees show the success of John's experiment.

Not alone in the lab

The image of a scientist working alone in an isolated laboratory in a castle or a basement is hard to shake, thanks to movies and television. John's laboratory is much more typical. "There are about ten of us in the lab at once," he notes. "Each person may be working on a different project, but we talk to one another all the time. It's noisy, but it's a great way to share ideas." John comments, "It's a good thing we put our names on everything or I could end up with alfalfa instead of spruce trees."

Last, but not least

John spends the rest of his day at the word processor in his office. He's putting the last touches on a scientific paper, a short article that describes his latest work. "I write a paper each time I feel I have some results that should be shared with other scientists," John explains. "The paper is reviewed and published in a journal. I also send copies, called reprints, to anyone who asks."

Spreading the word

Communication is key to good science. John sees communication as also being the key to the world's future as well. "Poor countries simply can't keep up with new information. They can't afford the scientific journals that we subscribe to." This means that agriculturalists and scientists in countries in many parts of the world are not always able to

Increasing in vitro germination of _Musa balbisiana_ seed

J.C. Afele[1] & E. De Langhe
Laboratory of Crop Physiology and Tropical Crop Husbandry, Katholieke Universiteit van Leuven, B-3030 Leuven, Belgium ([1] present address: Horticultural Science Department, University of Guelph, Guelph, Ontario, Canada N1G 2W1)

Received 4 April 1990; accepted in revised form 13 May 1991

Key words: embryo, in vitro germination, _Musa balbisiana_

Abstract

Seeds of _Musa balbisiana_ were soaked in water for five days prior to excision of embryos. Embryos with their longitudinal axis laid flat and half-way embedded on agar-solidified medium produced the highest germination and the most desirable plantlet characteristics. Germination in vitro was 94% within 7 days compared to 50% after 54 days for greenhouse-sown seeds.

Introduction

Bananas form a major staple in most African countries and are a favourite fruit in the developed world. The wild progenitors of the edible bananas (_Musa acuminata_ and _Musa balbisiana_) produce seeds freely while most of the edible clones are seedless, with a few notable exceptions such as 'Pisang Awak' (subgroup ABB) (Simmonds 1966). The creation of new cultivars through breeding is hindered by low seed set, as well as slow and nonuniform germination (Shepherd 1954, 1960; Stotzky et al. 1962).

The difficulty of obtaining seed for breeding has increased interest in in vitro germination of both intact seeds and excised zygotic embryos (Cox 1960; Stotzky & Cox 1962; Stotzky et al. 1962). Germination of excised embryos is influenced mainly by two factors: the maturity of the embryo at excision and composition of the culture medium (John & Rao 1984). Orientation of explants on solid media has been shown to influence the developmental pathway of explant tissues or organs in other species. For instance, anthers of rice (Mercy & Zapata 1987), and

barley (Shannon et al. 1985), plated edge-wise with only one lobe in contact with medium, produced more embryoids capable of plantlet differentiation than those plated flat with both lobes in contact with medium.

Preliminary investigations with banana seeds indicated that soaking intact seeds prior to embryo excision and culture might improve germination. The object of this research was to determine the effects of soaking intact seeds before embryo isolation, and embryo orientation on agar-solidified media on germination.

Materials and methods

Embryos of _Musa balbisiana_ diploid seeds, produced by open pollination and received from Honduras, were soaked in water for various time periods from 0 to 9 days, washed sequentially in 70% ethanol for 2 min, then 1.5% sodium hypochlorite solution for 20 min and rinsed three times in sterile distilled water. A longitudinal fissure was made in each seed and the whitish, mushroom-shaped, embryo was removed and cultured on modified semi-solid Knudson

This is an article published in a scientific journal. To help other scientists locate this information, the title and a summary (called an abstract) are listed in computer databases as well.

receive the latest information published on biotechnology — information John feels would help them immensely. But he has found a way to help.

"I've started collecting practical information on plant biotechnology from researchers all over the world, especially information about tropical plants. I plan to produce a newsletter that combines this information in a useful form." John will send this newsletter to researchers in countries that need it.

"There is so much promise to this technology," John says earnestly. "It can help solve food shortages and economic problems that are now crippling whole nations. It must reach the people who need it and can use it."

Before he leaves the laboratory, John takes a moment to plan the next day. "When I arrive in the morning, I want to know exactly what needs to be done. I always want to be moving and thinking ahead," John smiles. "It's my nature to set goals for myself."

Activity

Cloning a carrot

Plant scientists such as John Afele know that parts of a plant can be used to produce clones — identical copies of the original plant. You can perform this experiment to do some cloning yourself.

You will need
fresh carrot (not cooked or frozen)
sharp paring knife
shallow pan or pie plate
plastic wrap

Procedure
Arrange your carrot slices in the pan, and label the pan in a way that allows you to check from which portion of the plant each slice came.
1. Fill a shallow pan with water to a depth of about 1 cm.
2. On a cutting board or a clean dish, carefully cut a carrot into pieces, as shown in the diagram.

cut at about 2 cm intervals

3. Place each piece in the pan so that it is half-submerged in water.

remove any leaves

4. Cover the pan with the plastic wrap and wait for two days. (Do not let the water dry up!)

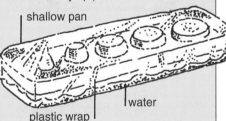

shallow pan

plastic wrap | water

What happened?
After three days have passed, look for signs of successful cloning — new leaves and roots. Which part or parts of the carrot plant can be cloned?

Challenge

Repeat this experiment with other vegetables, such as a parsnip or a potato.

How to become a plant scientist

The first step in becoming a research scientist in any field is to attend college or university. During the final years before they graduate, most students specialize in a particular kind of science. For example, John specialized in crop science. "I found I liked science that gave me a practical goal."

Exchange programs

After graduation, John took advantage of an agricultural exchange program offered at his university in Ghana, Africa. This program matched students in Ghana with students in Europe. "We arranged summer jobs for each other, so we could exchange firsthand knowledge about farming." John found a farm job in Ghana for a Belgian student. "And this student found me a great farmer to work with in Belgium. The farmer is still a good friend."

Advanced studies

The next step is graduate work. This involves some course work, but consists mostly of research done under the guidance of a professor. John wanted to do graduate studies in plant biotechnology. "I was accepted by two universities that were doing the work I was interested in, one in Canada and one in Belgium." Since John was already living in Belgium, he decided to do graduate work there.

Going bananas

"When I tell people I worked with banana seeds, they're usually surprised. Although edible bananas do not have seeds, non-edible bananas do produce seeds. "Normally, banana seeds take 60 days to germinate. So farmers prefer to use underground stem cuttings to grow new plants. But then all the plants grown are identical. You lose the variety you get by combining cells from two different plants to produce seeds. This means that if a disease were to strike, all the plants might die because they're all identical." John's research led to the development of an important new technique. By the end of his project, John was able to get banana seeds to germinate in just two days. "A lot of researchers are now using this technique in their work."

Lifelong learning

When John finished his Master's degree, he went to Canada to work on his doctorate. "My Ph.D. project was on corn. Since graduating, I have done several other projects at the university." Each project has given John more tools to use and more knowledge to guide his future research. "Everything a scientist does is based on what he or she did before," John notes. "A scientific person is curious. You seek answers and information all the time. That's how you develop your skills."

Is this career for you?

People who will become good scientists are not satisfied with vague or incomplete answers. They have a drive to understand the world around them. After all, questions that can't be answered easily are the questions that lead to research.

If you were to become a plant scientist, your research could be on anything about plants that interests you. Or — as in John's case — your projects could be based on a need to help solve a problem such as world hunger. The possibilities are almost limitless. "A friend told me I'm ambitious, which surprised me," John recalls. "I've never thought of myself that way. I simply have goals that matter to me and I intend to achieve them."

Scientists must be keen observers. It helps to be a person who enjoys and pays attention to details. John has noticed something else in common among the scientists he knows. "They don't get discouraged," he laughs. "Frustrated maybe — but only in a way that inspires them to try harder."

If you decide you'd like to be a plant scientist who studies plants in the wild, you might have to put up with conditions like this. These two plant ecologists are looking for rare rain forest species in Australia's Arnhem Land.

Career planning

Start a curiosity notebook. Whenever you are curious about something, record the topic in the form of a question. After a few days, reread your questions and see which ones interest you most. Find answers to these.

Making Career Connections

Improve your organizational skills by keeping a weekly study schedule — and sticking to it. In your schedule, plan time for extra reading as well as homework.

Contact a horticultural or agricultural association, private or government, for information about crop science and plant biotechnology.

Write to the college or university nearest you and ask for information about the science programs they offer. Your school guidance counselor can help you, or you can write directly to a nearby college's science department.

Getting started

Interested in being a plant scientist? Here's what you can do now.

1. Plan on taking courses in science, especially biology and chemistry, in high school. Don't forget math.
2. Look for summer or part-time jobs working with plants. Practical knowledge is always helpful.
3. Find out about financial aid for university, such as scholarships or grants. Many of these should be applied for while you are in high school. Find out about exchange programs, too.
4. Contact the college or university nearest you and arrange to visit the botany or horticulture department. Ask if there are any opportunities for research assistants.
5. Join a science club. If there isn't one at your school or library, start one.
6. Find out if there are summer science camps offered near where you live.

Related careers

Here are some related careers you may want to check out.

Horticulturist
Usually works with cultivated plants that are not primarily grown as food crops. Duties may range from supervising the production and growth of plants to developing new varieties.

Plant ecologist
Investigates plants in their natural ecosystems. Sometimes models ecosystems in the laboratory. Provides information on how to preserve and maintain natural environments.

Crop specialist
Stays informed of new advances in crop science and plant biotechnology. Provides advice to farmers on how best to care for particular crops, as well as what variety of crops to grow. May travel to other countries to exchange information.

Plant geneticist
Develops new or improved varieties of plants using traditional or biological methods. Researches how plants evolve and the relationships among different plant species.

Future watch

"Biotechnology is important in every field of science," John remarks. "There are more openings all the time for researchers with this specialty." This isn't just happening in colleges and universities. Many new "biotech" companies also apply the technology to practical problems.

Fahmida Rafi — Museum Biologist

Fahmida Rafi collects, identifies, and studies organisms called isopods. "These fascinating creatures are crustaceans. Their closest relatives are animals like crayfish and lobsters," Fahmida explains. Although most isopods live in water, you are probably only

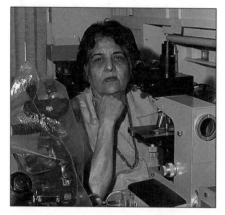

familiar with one that lives on land — the pill bug (also called the wood louse, or sow bug). These are creatures with seven pairs of legs that scurry away when you lift a log or a pile of leaves.

Sometimes Fahmida and her colleagues are at sea for days, collecting samples of marine life. The information the researchers obtain will be vital to understanding fragile ocean ecosystems. But at this point, all they have are jars containing a bewildering array of tiny organisms. "The first step is to 'fix' the organisms using a substance that leaves them as much as possible in the form in which we found them. Then we preserve the organisms so they can be examined later," says Fahmida. "The second step is to label the jars with all the important information about where we found the samples."

Fahmida sorts out the major groups of organisms, such as

sponges, worms, crustaceans, and molluscs. Then she concentrates on the isopods. "I give the other kinds of organisms to other specialists to identify. I rely on their knowledge and they rely on mine." By finding out what other animals were collected with the isopods, Fahmida can better understand the ecology of the organisms.

Naming the unknown

Back at the large museum where she works, Fahmida begins to identify what she has collected. It's delicate work, mostly using a microscope. What is especially exciting is discovering new species. "The first specimen of a new species found and described is called a holotype," Fahmida explains. "It will be used for comparison with any other specimen we think could belong to the same species." Holotypes are kept under lock and key and are never allowed to leave the museum. This doesn't mean they are unavailable. "Scientists and students from all over the world travel to the museum to study the specimens," Fahmida says.

Most isopods are smaller than a fingernail, but this deep-sea species is more than 35 cm long. Isopods are common and widespread, yet there is a lot biologists don't know about them.

Computer links

"Every time a specimen is donated to the museum, whether by the public, by researchers, or by our staff, it is recorded in the computer's database. This database is a treasurehouse of information, gathered piece by piece from around the world," Fahmida notes. It also includes information on the organism's habitat, such as the vegetation, the water level, or the type of place in which it lives. "Scientists in other cities can access this database from their computers," Fahmida points out. Thus, database information is available for many kinds of research, from fisheries to pollution studies."

A collection specialist must have a Master's degree in biology or zoology (the study of animals). "It's a great job," smiles Fahmida. "It takes me to many interesting parts of the world. It also allows me to do what I love most — learn about animals and share what I've learned."

Getting started

1. Take science — especially biology — and math in high school.
2. Ask if your local high school offers a co-op program in biology with a local university, museum, or college.
3. Keep a nature log, a notebook in which you can practice taking detailed notes about what you observe. Record-keeping is a vital part of scientific research.
4. Become familiar with computers and computer databases.

Ron Priem — Arborist

An arborist looks after trees in towns and cities. Why do trees need care? "Trees in cities aren't in a natural environment," Ron explains. "They are stressed by pollution and may not have enough space to grow." Such trees are more likely to suffer from disease or damage by insects.

Fitness for trees

Ron Priem's job is to keep trees healthy. When he checks a tree, Ron first looks for weak branches. These must be cut off, or pruned, before they break and damage the bark. The bark is the tree's main protection against insect pests and disease-causing organisms.

"I was happy to be able to save a 200-year-old red oak tree recently," Ron says. "It had a large hole in its trunk. By filling the cavity with cement, I was able to repair the tree and it's now doing well." Trees like this are rare and worth saving. "We admire the beauty of trees in our cities," Ron explains. "And they improve the environment by cleaning the air." Perhaps even more important, recent studies show clearly that the presence of trees around buildings helps moderate the temperature. In summer, less air conditioning is needed to keep buildings cool, and in winter, less heating is needed to keep buildings warm. Trees thus help conserve energy.

When nothing can be done

Sometimes a tree is too diseased to survive, or has been damaged by lightning. Then Ron is called in to remove it. "I bring a crew of six other people with me," Ron says. "I work

Ron needs over 300 m of rope and reliable safety equipment to work high up in large trees.

By cutting off limbs that are weak or that interfere with other limbs, Ron helps keep the entire tree healthy.

up in the tree, cutting from the top down. The crew cuts up the fallen branches and keeps the work site clean." Working high in a tree calls for strict attention to details. "It's critical that all safety equipment is checked," Ron says. "My ability to concentrate is equally important."

Looking for answers

Tree care is a fairly new field. There's always more to learn. When Ron encounters a tree that has a disease he doesn't know, he checks through reference books. "If I need more help, I'll take a small branch with leaves from the tree to the local university. The plant scientists there are glad to help."

New cells will grow over the cut area, healing it much as a scab is formed by human skin. This photo shows a healed trunk from which a branch was cut some time ago.

Getting started

1. Be aware of trees. Find out about different kinds of trees and observe trees in your community during each season.
2. Grow your own shrubs and trees.
3. Try to obtain a summer job with an arborist to see if you like the work. If the arborist can't pay you, consider volunteering. What you learn may soon lead to a paying job.
4. Write for information to a group that is involved with trees. For example, the International Arboculture Society provides courses and licenses for arborists.

Jill Cherry — Horticulturist

"**A**nyone who loves plants should consider a career in horticulture," says horticulturist Jill Cherry. (The Latin word "horti" means, quite simply, garden.) Jill coordinates plant production for a large city. To pursue this career, she completed a two-year college program in horticultural technology.

"We grow all the annual plants, such as petunias, impatiens, and geraniums, that will be planted on city property," Jill explains. "This year, that worked out to 600 000 individual plants." Jill supervises a staff of 20 full-time gardeners and growers. "My staff and I work year-round. There's always plenty to do."

Jill and her staff are also responsible for providing indoor plants for the city's offices, including City Hall. "The office environment is dry and warm," Jill notes. "So we grow plants that can survive these conditions."

Besides her work in the greenhouse, Jill serves on many committees, sharing information with other horticulturists. "We still need to learn about plants, especially plants in greenhouses and city parks. So I meet often with people from other cities to share ideas."

How it's done

The plants are grown in the city's huge production greenhouse, which has all the latest technology. For example, the automatic watering system is controlled by computers. "We also use computers to control the temperature and the humidity in the greenhouse," Jill observes. "This means that with very little human effort, our plants are kept in the ideal environment for growth.

"I try to avoid using chemical pesticides," Jill explains. "Whenever possible, we use the natural enemies of the pests instead." For example, tiny parasitic mites are released to combat certain insect pests in the greenhouse. "The mites destroy most of the insects, and don't harm the plants. It's a good way to solve the problem."

Special tasks

Jill's duties include something unusual: maintaining the plants of a 90-year-old historic conservatory. "The conservatory is a glass-roofed building where people can admire

The greenhouse that Jill operates. Jill and her staff use computer printouts that allow them to keep track of exact conditions in each part of the greenhouse.

exotic plants such as orchids, even in winter. One plant, a palm tree, has been growing in the conservatory since it was built. The tree has survived some unbelievable changes in this city." Jill pauses. "But when you think about it, perhaps the only thing that hasn't changed is our love for plants."

Jill discusses greenhouse techniques with two members of her staff. They, too, have training in horticultural technology.

Getting started

1. Learn about horticulture firsthand by growing your own plants. Helpful information is available from libraries, garden clubs, and nurseries. In some cities, you can get permission to have your own garden on public property.
2. Work in a nursery. A large part of horticulture is learning from others, especially from growers with years of experience.
3. If you have access to plant lights, try growing your own annual plants from seeds. If you are successful, you may want to start a small business or support a local charity by selling annual plants in your neighborhood.

Gino Maulucci — Scientific Illustrator

Do you know someone with a mind for detail, an interest in science and living things, and a talent for drawing terrific cartoons? Maybe it's you. If so, you might enjoy the career chosen by Gino Maulucci, an artist who draws and paints living things.

"I received a degree in biology, after studying primarily zoology," Gino recalls. "I wanted to do something that involved animals, or maybe medicine." When Gino graduated, he didn't have a clear idea of what he wanted to do for a living. "All the time, though, I was drawing for fun. So I took a summer course in art while I thought about my future." During this art course, Gino discovered that he had the talent and the knowledge needed for scientific illustration. He enrolled immediately in a special program to learn more about medical illustration. "Since then, I've worked for myself, and I love it."

Tools of the art trade

Gino draws for scientists, physicians, and science publishers. He uses any medium (art material) that suits the needs of his client. "For example, I might be asked to produce slides or

an animated video instead of a line drawing. Or I might create a color poster." Regardless of the type of art, Gino's most important tool is what's inside his head. "I depend on what

"I draw science-based cartoons for magazines read by physicians and dentists."

This picture shows a small part of a marine food web Gino recently painted. To paint the food web, Gino has studied the body shapes and the patterns of movement for each animal, and he has carefully considered where to place each animal. "I had to keep in mind the scientific principles being illustrated," Gino recalls, "but I wanted good artistic value, too. Without good art, the animals wouldn't be interesting or believable." An overlay will provide the arrows between organisms to show feeding patterns.

I've learned about science to help my art. If I need more information for a piece of art, I go to my collection of photographs. Working from photos helps me get the proportions right the first time." If Gino still needs information, he visits a local museum or university. "I often use specimens in order to draw color and textures correctly."

The first thing Gino does is a quick sketch, called rough art, to show the client approximately what the final art will look like. "For animation," he notes, "I do storyboards to show the basic steps in the action." Once the client approves the rough art, Gino prepares the final art in whatever form the client requested.

Laughter is good medicine

Physicians and dentists are a tough audience, because the humor works only if the science and art are perfect. "I enjoy the challenge — and the laughs. That's what's great about this job. I can put my own personality into everything I do."

Getting started

1. Take science, art, and math courses in high school.
2. Draw all the time, draw everything, and then draw some more.
3. Learn how computers are used in art and in publishing.

Classified Advertising

HELP WANTED

COMPUTER OPERATIONS SUPERVISOR

Retail chain with P.O.S. (DOS), back office system (UNIX), accounting experience. Debra 444-4444

SENIOR Computer Artist needed as a Software Operator. Must have a minimum of 18 months' experience with similar software. 35 hr. work week, rotating shift work involved. Salary $36,400/yr. Please forward resumé to Box 2040, The Daily News, City, Province/State.

UNIX/C Programmer. Some international travel. Please call 647-2105.

LABORATORY Technician for time research position at local hospital. B.Sc. or M.Sc. and experience in molecular biology required. Call 357-1492.

PHYSIOTHERAPIST

The Community Health Center is seeking a part-time therapist. This health center offers a multidisciplinary approach with special focus on Senior Education, and Health Promotion. Competitive salary with excellent benefits. Candidates must be eligible for registration with the Society of Physiotherapists.

Please send resumé by October 16, 19— Community Health Center 30 Smith St., Unit 201, City, Province/State; Attention: Executive Coordinator

KEYBOARD DEPT. MANAGER

Responsible for inventory management sales and marketing. Must be hardworking, with sales experience and product knowledge in some of the following: Home Keyboard, Pro Synthesis Home Recording, and Music Software. Sales Position also available. Forward resumé to: **MusicComp, 392 East St., City, Province/State.**

Electronics Engineer

Traffic Systems Company Limited requires an Electronics Engineer for a successful...

Manager, Environmental Services

Fuels, Inc. is seeking a bright, self-directed professional with a broad understanding of environmental issues to assist in the development and coordination of company environmental policies and programs.

...include participation in environmental assessments of new facility construction, review of operations procedures, waste management strategies, advising company departments and regions with respect to compliance with environmental legislation, environmental auditing, and monitoring. Internal research and government relations... Inc.

...minimum of 7 years... familiarity with... distributor. Strong... competitive salary and... consideration, ...Inc.

> **WANTED**
> ## Kennel Assistant
>
> The Laclie Emergency Veterinary Clinic requires a part-time Kennel Assistant for rotating shift hours, including some weekends and evenings. The Kennel Assistant will work under the direction of our veterinarians and our Kennel Supervisor. Tasks are to look after stockroom inventories, assist in the handling of animals, clean the cages and floors, and exercise the animals. Applicants must have experience working with animals and must demonstrate a responsible attitude.
>
> **Submit your application to the Clinic at 650 River Road, Vaughn, Province/State, Postal/Zip Code, attention Dr. Helena Richards. (No phone calls please.)**

...of Mechanical Engineering Tech-op Ltd. 402 Maple St., City, Province/State (Tel. 7182; Fax 24...)

...all applicants for applying. However, only those under consideration will be contacted.

...the distribution of electronic security/fire/safety systems. We are currently seeking an individual to sell systems. This position will be based our corporate office and will require approximately 20% field travel. Qualifications include a sales background in the electronic security/fire industry and strong interpersonal skills. For consideration and a local interview, applicants should forward resumés, complete with salary history, to: **Mr. R. Treed Security Inc. 319 Southfield Drive City, Province/State**

Position Available

An inner-city Community Health Center, with community and clinical activities, is looking for experienced following positions:

Nurse Practitioner: B.Sc.N. or equivalent experience in community-based primary health care and program planning. Strong clinical skills required.

Health Promoter: Health promotion degree or equivalent. Demonstrated success in developing and implementing health promotion programs at a community level. Good communication and group facilitation skills required.

Community Health Worker: Social services degree or equivalent. Demonstrated success in developing and implementing community-based projects. Grassroots advocacy and community organizing experience an asset. Familiarity with community health issues and resources a must.

Successful candidates will have:
- experience with any of the following: street people, ex-psychiatric patients, low-income people, families in crisis, and/or immigrants.
- the ability to work well both independently, and as part of a multidisciplinary team.
- multicultural experience and/or knowledge of other languages

Send your reply, indicating the position you wish to apply for, to: Carol Klim, Program Coordinator, Central Community Health Center; 3 Augusta Avenue, City, Province/State; Fax 363-2115.

Pharmacist

You will be responsible for delivery of all inpatient, outpatient, and retail pharmacy services. This position requires a professional designation, and a minimum of 2 years' related experience within a retail or hospital pharmacy operation. Excellent communication skills and the ability to work independently are essential to your success in this job. The location will appeal to individuals who enjoy extensive outdoor recreation activities, including kayaking, boating, exploring many small islands, and fishing. Along with this wonderful, close-to-nature environment, the successful candidate will enjoy a competitive compensation package and a subsidized housing package. Qualified candidates are invited to apply to:

Administrator, R.W. Large Memorial Hospital City, Province/State, Postal Code/Zip Code; Tel: (000) 357-2314, Fax: (000) 357-2315

Who got the job?

Finding a job

The first step to success in any career is getting a job. But how do you go about finding one?

- Talk with family, friends, and neighbors and let them know what jobs interest you.

- Respond to "Help Wanted" ads in newspapers.

- Post an advertisement of your skills on a community bulletin board.

- Register at government employment offices or private employment agencies.
- Contact potential employers by phone or in person.

- Send out inquiry letters to companies and follow up with phone calls.

A job application usually consists of a letter and a resumé (a summary of your experience and qualifications for the job). Applicants whose resumés show they are qualified may be invited to a job interview.

Activity

Hiring a kennel assistant

The advertisement shown on the opposite page was placed in a local newspaper. What do you think it takes to be a good kennel assistant?

For the position of kennel assistant, there might be many applicants. The letters and resumés submitted by two applicants — Maria Acieto and Don Leduc — are shown on pages 46 and 47. Read these letters and resumés, and make notes about whether each applicant might qualify for the job.

Maria Acieto and Don Leduc were granted interviews. Dr. Helena Richards, one of the veterinarians at the clinic, conducted the interviews and made notes about each applicant. Her notes are also shown on pages 46 and 47.

Procedure

Use the letters, the resumés, your own notes, and the interviewer's notes to list the strengths and weaknesses of each applicant.

If you were Dr. Richards, which person would you hire: Maria Acieto or Don Leduc? Perhaps you feel that more experience is necessary for the job; you might decide to hire neither applicant.

Challenge

How would you perform in a job interview? Role-play an interview with a friend. Ask your friend to play the part of a veterinarian interviewing you for this job. Then reverse roles. Role-playing will give you practice asking and answering questions. This practice can help make sure that when you apply for a job, you have a good chance of getting it!

Maria Acieto's application and interview

1316 88th Street
Vaughn, Province/State
Postal/Zip Code

September 15, 19 —

Dr. Helena Richards
Laclie Emergency Veterinary Clinic
650 River Road
Vaughn, Province/State
Postal/Zip Code

Dear Dr. Richards:

I wish to apply for the job you advertised in The Sentinel on Saturday, September 13, 19—. I am enclosing my resumé for your consideration.

I love animals and I have had a lot of experience with them. I have two cats and a dog, and I used to raise rabbits. I take care of all my pets myself. When our neighbors leave on vacation, I look after their two dogs as well. I exercise and feed the dogs and clean out their kennels.

I am a very responsible person. For the past two summers I have helped my parents tend our vegetable garden and flowerbeds.

I like being busy and working hard. I feel that I can do the job of Kennel Assistant at your clinic. I hope you will ask me to come in for an interview. You can call me any time after 3:30 p.m. at 555-1096.

Sincerely yours,

Maria Acieto

Maria Acieto

Resumé

Maria Acieto
1316 88th Street
Vaughn, Province/State
Postal/Zip Code
Telephone: 555-1096

Age: 16

Education
I'm taking a computer course as well as my regular courses.

Interests and Activities
I am in the Camera Club and play on my school's volleyball team. My hobbies are animals and old movies.

Job Experience
Babysitting: Evening and summer babysitting service; responsible for the care of children aged eight months to seven years.

Pet Watching: Cared for two purebred huskies for three weeks, including feeding, exercise, brushing, and cleaning out kennels.

References
Mr. Alek Jakalski
1320 88th Street
Vaughn, Province/State
Telephone: 555-2018

Ms. Cristina Fernandes
Volleyball Coach
Vaughn High School
192 Allstate Road
Vaughn, Province/State
Telephone: 555-5564

Interview: Maria Acieto
• Maria arrived on time for her interview; wore a blouse, jeans, and sandals.

Why does she want the job?

• She loves animals, has always wanted to be a vet, and wants to see what the job is really like.

What would she do with a snarling dog?

• She'd ask the owner to handle it.

Can she work holidays and late evenings?

• Holidays fine, but can't work later than 9am on a school night.

HR

Don Leduc's application and interview

33 Michigan Road
Apartment 15
Vaughn, Province/State
Postal/Zip Code

tember 16, 19 –

Helena Richards
clie Emergency Veterinary Clinic
50 River Road
aughn, Province/State
ostal/Zip Code

Dear Dr. Richards:

I'm the person you've been looking for to be your new Kennel Assistant! As you can see from my resumé, I have experience working with both large and small animals.

I want to be a biologist and do research on wolves. Being a Kennel Assistant would really help me in my future career. I would like to meet you and tell you more about why I would be the right choice for this job. Please contact me by telephone at 555-2372, or by writing to the above address. Thank you for your consideration.

Yours truly,

Don Leduc

Don Leduc

Interview: Don Leduc

• Don was early for his interview; wore a shirt and tie.
Why does he want the job?

• He wants experience, as wildlife biologists have to be able to understand and handle animals.
What would he do with a snarling dog?

• He isn't afraid of dogs and wouldn't take any nonsense.
Can he work holidays and late evenings?

• He can't work some holidays because of family commitments, but late hours are okay.
HR

Resumé

Donald R. Leduc
33 Michigan Road
Apartment 15
Vaughan, Province/State
Postal/Zip Code
555-2372

Age: 17

Work Experience
Assistant Ranger: I have spent three summers working as a volunteer ranger in Asseqy National Park. My duties included catching and banding birds, helping to make trails and campsites, and helping the park biologist prepare exhibits for the public.

Hobbies and Interests
I am involved in Scouts and help out with Cubs. I like most sports, especially outdoor ones.

Education
I will graduate from Vaughn High School this year. My best subjects are English and science. I have taken two First Aid courses and have my Lifesaver's Badge in swimming.
References available on request.

Index

Credits

(l = left; r = right; t = top; b = bottom; c = center; bl = bottom left; br = bottom right)

All photographs by Roger Czerneda, Crystal Image Photography, except 6(bl) R.R. Sallows Collection, Ontario Ministry of Agriculture and Food; 6(br) Canadian Therapeutic Riding Association; 11(r) James D. Rising; 14 Brian Hicks/ Metropolitan Toronto Police; 16 Dr. Karen Goodrowe; 17(l and r) Dr. Karen Goodrowe; 18(t and b) Dr. Karen Goodrowe; 19(t) Catherine Rimmi; 19(b) James D. Rising; 20 Dr. Karen Goodrowe; 25 James D. Rising; 31 MSDAGVET, a Division of Merck Frosst Canada Inc.; 32 Karen Machin; 37(t) John Afele; 38 James D. Rising; 40(r) Len Marhue, Canadian Museum of Nature; 42 Catherine Rimmi; 43(b) Gino Maulucci art photographed by David Rising; 43(t and c) Gino Maulucci.